Forest H. Belt's

Easi-Guide to CB RADIO FOR THE FAMILY

Text and Photography by

Marti McPherson and Forest H. Belt

(Mother Hubbard and Easi-Reader)

HOWARD W. SAMS & CO., INC.
THE BOBBS-MERRILL CO., INC.
INDIANAPOLIS • KANSAS CITY • NEW YORK

FIRST EDITION

SECOND PRINTING—1976

International Standard Book Number: 0-672-21330-3
Library of Congress Catalog Card Number: 75-36963

Preface

Some families have a knack for playing together and staying together. As often as not, part of the reason includes some hobby they hold in common—an activity each individual enjoys, yet all can join in collectively. Citizens Band radio rates as one such pastime.

CB radio has in fact exploded into one of today's most popular electronic interests. Television and stereo music hold the number-one and -two spots; almost every family has both. Yet, as I write this, a fascination with personal two-way radio communications, which is what Citizens Band radio amounts to, has begun sweeping family after family.

Two-way radio is not new for police, fire, and other public safety outfits, nor for taxicabs, boats, and airplanes. But it's expensive. And regular two-way radio laws make no provision for the ordinary citizen. Until CB radio, your only home-to-car opportunity involved costly telephone-company facilities. Citizens Band radio has opened a wide vista of new communication possibilities. Two factors bring this treasure within reach of almost anyone: the equipment costs comparatively little, and the license for operation is cheap and easy to obtain. Those circumstances spell popularity, and CB radio has been gaining headlong in popularity.

CB radio originally was not intended for a hobby. It was for convenience, so anyone could have quick personal access—via two-way radio—to their business employees, or to family members, while on the move around town. That was the plan in 1958, when class-D Citizens Radio was instituted. Talking to friends was permitted, but under limited circumstances.

The idea grew so rapidly, though, that people began using CB for all sorts of conversations. Soon almost every CB owner was chatting with anyone he could find "on the channels," though not legally. In early 1975 the FCC revised the Rules to permit CB radio operation as a hobby too.

Today, *YOU and your family* can enjoy CB radio to your hearts' content. CB radio can save you time around town. On the highway, as you'll discover, CB can save your life. You can form lasting friendships near home, or in new localities when you travel. It is to you who want and need this family-type of CB radio enjoyment and utility that writer/photographer Marti McPherson and I dedicate this book.

Naturally, we use CB radio ourselves, for both pleasure and business. We have met a lot of friends on the CB channels, some of whom we never met in person. Some of these CB acquaintances helped in various ways with the preparation of this book and its photography.

Our gratitude goes to Greg Thomas of New/Era Sales, Russell Ott of Comtronix, Jerry Woodward of Jay's CB Radio Shop, Big R's CB Radio Shop, Creston Village Radio Shack, Alger Camping Supply Co., Debbie and Steve McPherson, Bruce Jensen, Scott Simons, and a few South Country (Indianapolis area) CBers: Lady Ragtop and Ragtop, Gadabout and Lady Gadabout, Jenny-Bird, Gorilla, Tinkerbell, Cricket, Boogie Man, and Rollbar. We appreciate their unselfish efforts.

Several companies involved themselves too. We have included photographs from Browning Laboratories (page 73), Cush Craft (36), GC Electronics (33, 67, 72), E. F. Johnson (30), Kustom Kreations (89), Magitran Corp. (109), New-Tronics (Hustler) (93), Pathcom Corp. (Pace) (13, 28, 38, 79, 123, 124), Radio Shack (Realistic) (9, 50, 52), REACT National Headquarters (80, 81), Shure Brothers (85), and Teaberry (29). Certain firms contributed much more by loaning or supplying equipment for tryout and photography; we tender special thanks for that thoughtful help to New-Tronics (Hustler antennas), Anixter-Mark (Mark antennas), Antenna Specialists, Magitran Corp. (Poly Planar speakers), Dynascan Corp. (Cobra), E. F. Johnson, Kris Corp., Tram/Diamond Corp., and Linear Systems (SBE).

It's a measure of the high interest in CB radio that so many people wanted this book to serve as a real guide for every CBing family. They and we—Miss McPherson and I—wish you many pleasant CB gab sessions. And we hope this book does its job of making those on-the-air get-togethers easier and more worthwhile than they might otherwise have been.

Forest H. Belt

PUBLISHER'S NOTE: Other CB books by the author are *Forest H. Belt's Easi-Guide to Citizens Band Radio* and *Easi-Guide to CB Radio for Truckers.* He also authored *1-2-3-4 Servicing Transistor CB and Two-Way Radio,* a book on CB repairs and troubleshooting.

Contents

Chapter 1

Your Very Own Two-Way Radio

Almost everyone can talk now on their own two-way radios. From car to car, car to home, friend to friend, neighbor to neighbor—all over the country, folks chat together without the need for telephone lines. It's fun, and it's relaxing. They do it with Citizens Band radio.

Maybe you already know about *CB radio.* Certainly if you look, you'll see houses sprouting tall arrays of aluminum tubing (next page) that you know are not TV antennas. And you'll spot a rash of cars on the highways with one or two antennas poking up like stingers. And big trucks with slender rods attached to their side-mirrors. Once you start noticing, it seems nearly everyone must be using CB radio.

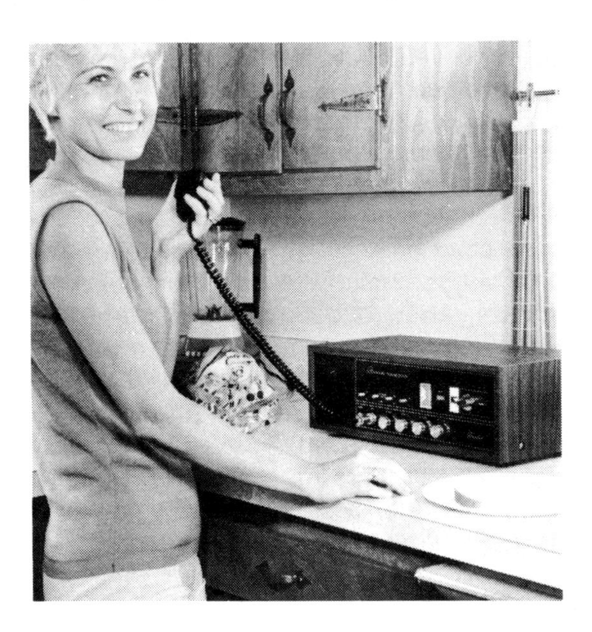

Okay, if seeing antennas on houses is your only contact, you may want to know just what the Citizens Radio Service is. The *Class D* segment comprises a "band" or group of radio frequencies between 26.965 and 27.255 million-cycles-per-second (called *megahertz* and abbreviated *MHz*). The Federal Communications Commission in 1958 set aside this band for public use.

Not many people took CB seriously at first. Ham radio operators, hobbyists who must pass a stringent technical exam for their license, scoffed at unknowledgeable CBers, who needed only to fill in an application. The FCC didn't take the Citizens Radio Service too seriously either. For one thing, CB wasn't intended as a hobby, like ham radio is. Furthermore, the radio frequencies set aside for class D Citizens Radio are fairly susceptible to interference; the band wasn't thought suitable for reliable communications.

How wrong they were. Now, so many years later, America has become CB-crazy, and for several good reasons.

First, over the years CB transceivers (transmitter-receiver in one unit) have been greatly improved and simplified. Modern CB outfits provide reliable communications on any of the 23 CB frequency *channels.* Sets are easier to use than a telephone.

Second, CB licenses are still easy to get. You don't need special skills as you would for ham radio (Amateur Radio Service). You don't need to know Morse code. And you don't have to pass any arduous technical tests to qualify for your license. You pay a fee, but it's only $4. With a license lasting five years, that's only 80¢ a year.

Third, CB radios are not expensive—at least not compared to other two-way radios, like police or taxicab outfits.

Your best use for CB radio involves situations where telephone communications are impractical. For example, on the road. CB radio has proved its usefulness to America-on-the-move to such an extent that a half-million Americans beat a path (by mail) to the FCC's door each year, applying for class-D Citizens Band licenses.

A few people cannot have a class-D Citizens Band license. A foreign government, an agent of a foreign government, or an alien living here cannot get a CB license. Nor can you hold a CB license unless you're 18 or older. (This may be reduced to 16 years old by the time you read this; the FCC is considering it.) Any other American citizen, operating as an individual, a small or large business, corporation, or association, or a department of Federal, state, or local government, can apply.

There are many pluses, where family CBing is concerned. Once an adult in a household has a CB license, he (or she) can authorize anyone living in the house to use the station. That extends to teenagers and younger children as well. But you, the licensee, are responsible for when and how and for what purposes your CB equipment is used. If you allow your children to use your CB equipment, don't turn them loose unattended until you have taught them properly how to use the equipment.

Make sure they know enough of the rules and regulations not to break any laws. Then teach them good CB manners (by example is one good way). Most CBers don't mind a youngster on the channels if the child doesn't abuse the privilege.

You'll discover that CB radio is an excellent way to involve your family in community happenings. Designate certain times your children can monitor the channels for emergency messages. You can develop in them a sense of responsibility at an early age.

Your CB transceiver comes in handy in a lot of places—your automobile, for one.

How many times has car trouble (summer or winter) stranded you on a lonely road? Since CB radio has grown so widely popular, you'll find someone listening in all but the most impossibly remote areas. Along highways, a service station within range of your signal may answer your call in seconds.

Say your car konks out one wintry morning on your way to work. As long as there's even a little juice in the battery, you can call for help. Here's how it might go:

Switch over to channel 9 (the emergency channel) and give a shout: "This is KEQ3427 (use your own call sign, of course), Mother Hubbard (use your own handle) with a 10-34 (trouble and need help). Do I have a copy?" The reply may come back, "Okay, Mother Hubbard, this is KZU0077, Blue Grease Monkey. What's your 20?" (10-20 is CB code for location.) Once you've told him where you are and the nature of the problem, he'll probably say how long it will take his service truck to reach you. Sure beats walking a mile or two, or hitch-hiking.

One thing about calling for help. Should you fail to draw an answer, try again. This time say you need emergency road service. If you still fail to raise anyone, switch to other channels and try. Hunt for one on which you hear a local conversation. Ask to "break" (see page 26), then state your difficulty. Someone might phone a service station to find you help.

If you drive a great deal, you'll feel more secure knowing you're not alone in a breakdown situation. When trouble arises—for you, or someone else—get on the CB and report the difficulty and the location. You'll be amazed at how quickly help arrives.

A *base* transceiver constitutes the heart of your family CB communication system. From that point you can keep in touch with your children, spouse, a sick family member across town, or your neighbor.

CB comes in handy when you've sent your youngster to the grocery and found you forgot to ask for something. Or when you want one of the family to pick up a loaf of bread on the way home. You go to your home transceiver, key the mike, identify your station, and call your mobile unit (in a car, usually). That "mobile unit" could even be a Citizens Band walkie-talkie, sometimes called a *hand-held* CB unit. Having established contact, pass on the information, and save someone a second trip.

If you have a base station at home and a mobile unit in the family car, you can usually find out within minutes what time your husband (or wife, or working son or daughter) will arrive home. You'll serve fewer cold dinners when someone is held up by traffic. Even if the person you're expecting is too far away to reach by CB, a CBer between the two of you can· relay a message. You might hear something like this: "KEW3427 base, this is KOU7865 with a message from Sprocket Ears." Hearing your call sign and a family handle will grab your attention, and you'll have the news in a matter of minutes.

You'll find that CBers in general are the friendliest people in any town. Someone's always willing to help a friend or a fellow CBer in need. If an ailing or elderly friend or family member has CB equipment, you can check periodically and know if and when help is needed. Some CB groups join forces to "look after" aged people in the community, dividing the time and chores of regularly calling those who have CB.

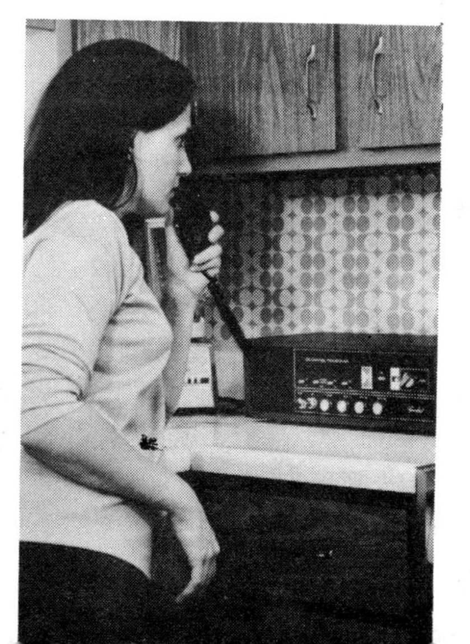

You can put CB two-way radio to work in the air and on the water, too. Many owners of private aircraft and small boats have already done so. Of course, marine and aviation radio serve you in your boat and plane, but conversations on those frequencies are limited to those relevant to navigation and operation. You can't use ship-to-shore radio to call your home directly, or to talk with someone waiting for you in a car at the dock. With CB radio you can.

If you have CB in your plane, you can radio home before you land, and ask someone to come pick you up at the airport. That eliminates leaving a car at the airport while you're out of town.

Also, CB on your boat or plane lets you chat (legally) with people in other boats and planes around you. You know how pilots like to swap flying experiences, and boat captains like to share a yarn or two. They're like CBers, that way. But they can't chatter idly on aircraft or marine frequencies. CB is the answer.

Because CB doesn't reach very far—only 10 or 15 miles—it's excellent for farm communications. In fact, in remote areas, it can seem almost like your own private wireless telephone.

No matter what the season, CB keeps you in touch with everyone who's working on your land. A "network" of farmers operating on the same CB channels can keep each other warned of approaching storms, flash floods, fires, even farm accidents. As long as there's a transceiver at both ends, you're in touch.

When you finish the day's outside work, you and your CB neighbors can chew the fat by radio about crop prices, livestock sales, stud services, and upcoming auctions or fairs, or you can plan next Sunday's barn-raising. The gals get together on CB to swap recipes, and make plans for canning and freezing. By radio they hear where the persimmons are ripe, or where to find some wild plums for jelly.

CB radio can save any farm dweller a lot of running around. On CB, several people likely hear you ask "Taterbug" if she knows where you can find some kumquats or rhubarb. If she can't tell you, someone else probably can. They'll ask to "break" into your conversation or they'll "shout back at you later" and give you the information. Neighborly helps like that may save you a dozen phone calls and hours of time.

CB in rural areas comes in handy as a locator device to help keep track of youngsters. If the kids haven't arrived yet from school, you can radio their friends' homes (if they have CB), or maybe even the school bus itself, and find out what's the delay. If teens have cars or trucks with CB, finding them is a little easier (when they don't turn it off to keep from hearing).

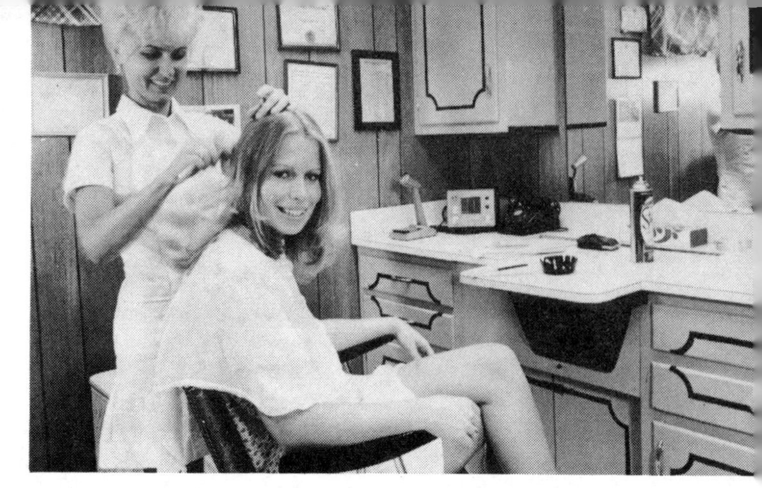

Besides all these uses of CB around the home, on the farm, on the highway, on the water, and in the air, think of its usefulness in your work.

If you're a carpenter or construction worker, you may be using CB on the job already. Unless you have a site office, phones are expensive and unhandy if not impossible. Even with a phone, you can't talk to the foreman or supervisor in stage 3 . . . or wherever. For the person who runs a small home-building company, a CB radio on-site (possibly a mobile) and one in your truck lets you track down the truck or ask if your helper found whatever you needed at the lumber company. You just grab up the mike and call him. You can give your foremen or superintendents new instructions throughout the day without their having to come to the office, or you having to go to the site.

You can easily imagine dozens of business uses. Perhaps, though, you never thought of a beauty shop having CB radio. It's a pretty smart idea. CB can find a place in a doctor's office, or anywhere that requires appointments. Traffic jams, road repairs (naturally begun the morning of your appointment), and 127-car trains can make you late. Your hairdresser can't know you've been driving around the block for 10 minutes looking for a parking place unless you call and say so. In the past, you couldn't. Today, with CB, you can.

People who run trade shows and conventions might find a base and several hand-held CB units indispensable. You could call a board or committee meeting on the spur of the moment; just give members a call on the CB. Sure, the PA system could be used, but that's only good as long as the people you want are all on the convention-center floor. And it's less private. You can even conduct a short conference on the CB, once you've made a rollcall. Some members might be in their cars enroute to or from the center. Others might be in an outdoor exhibit area. CB reaches them.

Obviously these are just a few limited examples of how you can use CB in business. Let these ideas stimulate ways for you to use CB in your own work. If your company has a large demand for two-way radio communications, you should investigate Business Radio. It costs more than CB, but you have your own almost-private channel—not so much sharing. But remember, you can't talk to other operators like you can with CB.

Remember when you used to drink gallons of coffee to keep awake on long drives? And can you still recall the exhaustion of your last driving "vacation"?

Well, take a tip from the professionals. One out of every five long-haul, over-the-road trucks has a CB radio. Listening to CB keeps drivers apprised of road and weather conditions. The chatter with and among other drivers keeps them alert far better and more safely than pills.

Smart travelers tune to CB channel 19 on the highways. That is the channel truckers use. This channel is not officially set aside by the FCC for use by truckers, but the channel has become popular through repeated use. Truckers swap up-to-the minute road-traffic advisories on this channel, and that makes driving safer.

Most people have heard of "Smokey reports." That's CB language for advice as to where police traffic patrols are roaming (or parked). Such reports are frowned on by some officers, and tolerated by others. More and more police find the reports help slow traffic down in dangerous reaches. For many reasons, highway travelers find "riding the rocking chair" (between two CB-equipped trucks a mile or two apart) a safe, efficient way to travel.

But Smokey reports are not all you hear on the trucker channel. You learn of accidents and traffic jams, and which lanes or roads help you avoid them. You find out about stretches of hazardous road, drunk drivers, or maybe where to find the best restaurant along a certain highway. And should you be driving late at night and running low on gas, truckers can tell you which stops are open and how far they are.

CB is extra fun when it's a family affair. It offers an attractive common ground for parent and child, and draws a family together naturally. CB radios are so easy to operate, any youngster can learn in no time to be a top-notch CB operator.

One advantage appears almost immediately. You're away from the house, but still in the neighborhood. You phone home, to a busy signal. (You know how teenagers are on the

telephone.) You'll discover it's far easier to "break" in on a CB channel than to cut into a telephone conversation.

Put your youngster to monitoring the highway emergency channel during homework. Motorists sometimes ask for assistance on channel 9. But it isn't often enough to prevent getting the book-work done. Show your kids how to give street directions, and teach them to assist motorists who need help. It's an unselfish public service, and can help develop personality, dependability, and responsibility.

CB may be a sometimes workhorse, but it should offer fun too. The FCC now permits hobby use of all CB channels except 9 and 11. But remember, the number of people using CB radio climbs every day. Exercise courtesy and caring when you operate your equipment. If none of us hog the channels or act foolish and childish on the air, we'll all be able to enjoy the Citizens Radio Service.

At one time Citizens Band radio was primarily citizens emergency two-way radio. It's still that and more. It's neighborhood news in action. Frequently, CBers get together for a "coffee," sometimes to collect money for another CB family struck by a sudden illness. Usually coffees are called only when the family's income stops, and the breadwinner is in the hospital for a serious operation. It's not charity, it's just a few friends helping out another.

Sometimes help is needed on a big scale. National CB groups like REACT (Radio Emergency Associated Citizens Teams) and regional outfits like the Kentucky Rescue Service step in with equipment and know-how to aid tornado, flood, earthquake, or hurricane victims. During such emergencies, you can help without even leaving your easy-chair. Just keep the channels clear of chatter for the time being. If you want to help more, stay close and monitor the stations involved *IN* the operation. You can pass messages along by telephone, by relaying on other CB channels, or by driving in your own vehicle to summon the necessary aid or run the needed errand.

One warning! CB radio is addictive. It grows on you—because it's people involving themselves with other people, getting to know their neighbors, and participating in bettering the community. Most CBers come to wonder how they ever got along without CB. That might happen to you.

Chapter 2

Talking CB

As you listen around on the CB channels, you'll notice two kinds of CB lingo. One is the offhand jargon that has evolved among avid CBers. If you can figure out what phrases like "come on," "threes to you," and "you gotcher ears on?" mean, you could probably skip some of this chapter. You'd learn that language from talking and listening.

Prepare yourself to talk with such characters as Under the Influence, Mother Hen, Mustard Muffin, Chipmunk, The Green Mango, Spiderman, Bird Dog, Yellow Flashlight, Circle City Vega Kid, Greasy Spoon, Hoot Owl . . . and the list goes on and on.

Then there's the technical talk. If words like linear, carrier, modulation, frequency, squelch, or sideband don't stop you cold—congratulations. You've probably been a CBer before.

CB jargon is a colorful language. The sooner you catch on to it, the sooner you'll become an accepted oldtimer. This chapter takes you from the tongue-tied beginner stages of CB grammar to, let's say, at least the high-school level. To acquire the real gift of smooth CB gab takes experience and practice.

Oh, by the way, "come on" is one of the friendliest, nicest-sounding phrases you'll hear in CBdom. It means, "Hey, it's your turn to talk."

All CBers must have a *handle.* It's not required by the FCC; they prefer the sterile, unimaginative device of three letters and four numbers, which is your *call sign* (see your license). But it's an unwritten law of CBdom that you take a handle. No handle, no conversation. Why is simple enough to understand. Handles are easier and more fun to remember than that cold, impersonal call sign.

Waterboy took his handle because he is always thirsty. A truck driver, he has the habit of drinking a couple glasses of water before he even tastes his coffee. A waitress once told him that if he ever became a CBer, his handle should be Waterboy. Well, he did, and it is.

Likewise, some handles infiltrate entire families. How about this family lineup: Cookie Monster, Cookiemaker, and Cookie Crumb.

Some CBers apply handles to their base stations. Pebbles and Bam Bam call their "home 20" the Bedrock Base.

So, if you've ever wanted to be someone else, here's your chance. You may never learn The Blue Goose's real name, even though you talk to him (apparently he doesn't know about ganders) several times a week. But you'll know him on the terms he wants to be known on. By the same token, other CBers will know you by your handle. Unlike legal names, handles can be changed easily. When you no longer like the image your handle projects, switch. No one will mind. They may just think they've found a new friend. And that's what family CBing is about anyway.

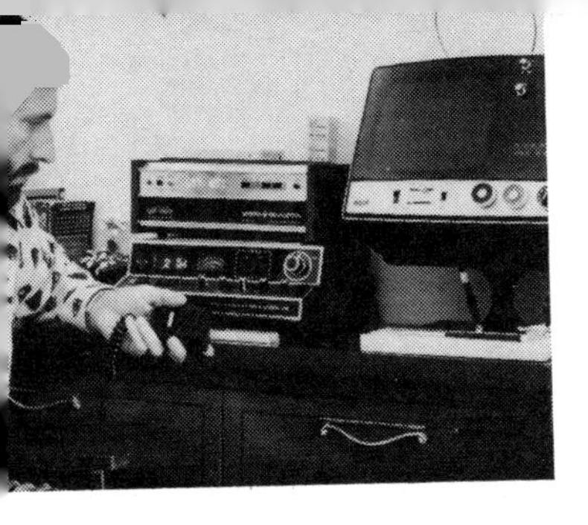

Okay . . . let's talk. All through this book we use snippets of CB conversation to initiate you into the world of CB. Here's one brief, typical two-way CB conversation.

I initiate the contact. "This is KEQ3427 mobile unit 1 calling base." I listen for a few seconds; no reply comes back, so I try again. (Maybe the base operator is in another room.) I key the mike and say, "This is KEQ3427 mobile unit 1 calling base. Hey, Easi-Reader, you got your ears on?" (In other words, is your CB unit on and can you hear me?)

Oh good; I get a reply. "This is KEQ3427 base, go ahead mobile 1."

"I'm on my way in. Do you have anything for me?"

"What's your twenty, Mother Hubbard?" (From 10-20, which means where are you?)

"I'm on Seventy-First, at Zionsville Road."

"Can you pick up a half-gallon of two-percent milk on your way?"

"10-4. Do you need anything else?"

"No. I —— —— . . . ."

"Breaker one-four" (Drat that breaker, he's overriding our conversation.)

"You just got walked on, Easi-Reader. What did you say?"

"All I said was, No that's all. I'll see you in a few minutes."

"Ten-four, Easi-Reader. This is Mother Hubbard in mobile unit 1, ten-seven at Kroger." (Going to be out of the car or otherwise out of reach by radio.)

"Okay, Mother Hubbard. This is Easi-Reader on the side for breakers. KEQ3427 base clear." (Will stay on the channel, listening only, to give others a chance to use it.)

It's a shame to have to say this, but leaving your CB transceiver visible in the car, even to go into the market for a quart of milk, is risky (see page 90). My 10-7 told Easi-Reader I was disconnecting my radio and locking it in the trunk. For more on the 10-codes, skip to pages 24 and 25.

When you're on the highway listening to a trucker channel, you'll hear that someone's "shaking the bushes" or "raking the leaves." If someone's already "minding the front door" and "back door," you can "ride the rocking chair." Here's what's going on.

The guy at the front door (ahead of you) is shaking the bushes (watching for road hazards and Smokey Bears). The guy tending the back door (behind you) is raking the leaves (making sure no bears come in). Your awareness is thus extended a few miles in both directions. If there's trouble ahead, you'll know in time to detour or adjust your driving speed. You'll also know of any trouble coming from behind—a speeding drunk driver, or a truck in the mountains with brake failure.

If you're not satisfied to sit in the rocking chair (between) and enjoy a free ride, at least remember to watch the milemarker signs (*mileposts,* some drivers call them). They're small green squares at every mile on interstate highways. Highway information won't much help unless you include the location of the trouble. You also report the lane direction in which the hazard exists. Also identify what road you're on; conversations from intersecting roads can lend confusion.

The following is a sample highway report. It conveys the necessary data without wasting time and words:

"KEQ3427 mobile unit 1 reporting a truck-car accident blocking the two right-hand lanes of eastbound I-70 halfway between milemarkers 116 and 117 (say one-one-six and one-one-seven). Traffic is beginning to back up, so slow down and move into the far left-hand lane. You can still get through there."

Because such highway reports provide a valuable service to all CB motorists—in trucks and four-wheelers alike—sensible CBers don't clutter the trucker channel with chatter. They'll appreciate it and so will you, particularly if someone spots a bear.

Smokey reports given over CB radio incite a lot of controversy. Smokey reports, if you haven't already guessed, is truckers' code for police activity along the highway. Some highway patrol officers have become hostile to CBers. Others maintain that Smokey reports actually aid in traffic control, not hinder it.

For example, a trucker reports "There's a bear with a camera in the center grove on two-eight-five at milepost one-five." He means a police radar car is sitting on the grassy median of interstate highway 285 at or near milemarker 15.

Another CBer may report a Tiajuana taxi eastbound on I-70 at mile 148 with lights on and the hammer down. He's telling you that a police car with a light on top is driving east at high speed on Interstate 70, with red-light flashing, and the CBer eyeballed (saw) him near milepost 148. If you're ahead of the police car, in the eastbound lanes, move over to the right-hand lane so he has a clear shot past you. If you're westbound, disregard the report.

One point further on Smokey reports. Sometimes you hear of a Smokey blackout. As you may know, truckers have proven very helpful to law enforcement officers in spotting stolen vehicles, drunk drivers, runaways, and criminals departing the scene of a crime. But with more people using CB, wrongdoers have caught onto the Smokey reports too.

Therefore, police with CB transceivers ask for a Smokey blackout. Truckers cooperate fully, and you should too. For the duration of the blackout, which may be a couple of hours or a few days, drivers refrain from reporting the whereabouts of the police cars in a given area.

Should you see a vehicle or persons matching the description of the wanted suspects, pull off at the nearest phone and call the police. The bears won't be too far away, and they'll be hungry. Don't try to detain criminals yourself, or you may be their next victim. And don't report them on the CB radio, or they'll be long gone when the troops arrive.

The 10-code is two-way-radio shorthand originally devised to speed and clarify communications on crowded police channels. If someone says "ten-twenty" (or just "20"), he wants to know your location. You'll hear "10-4" frequently; it means "message acknowledged."

But the CBer's 10-code differs somewhat from the police code. We have listed here the trucker's 10-code devised by the National Truck Stop Association, and the police 10-code. Others differ even more. There's also a 12-code, and an 8-code; you'll have to decipher those CBers' meanings as best you can.

*indicates complete difference in meaning

Number	CBer's Code	Police Code
10-0		Caution
10-1	Receiving poorly	Unable to copy, change location
10-2	Receiving well	Signals good
10-3	Stop transmitting	Stop transmitting
10-4	Okay, message received	Acknowledgment of message
10-5	Relay message	Relay
10-6	Busy, stand by	Busy, stand by unless urgent
10-7	Out of service, leaving the air	Out of service (give location and/or telephone number)
10-8	In service, listening	In service
10-9	Say your message again	Repeat
*10-10	Transmission completed, subject to call	Fight in progress
*10-11	Speak slower	Dog case
*10-12	Officials or visitors present	Stand by (wait)
10-13	Advise weather/road conditions	Weather and road report
*10-14	Correct time	Report of prowler
*10-15	We have passenger	Civil disturbance
*10-16	Pick up	Domestic trouble
*10-17	Urgent business	Meet complainant
*10-18	Anything (message) for us?	Complete assignment quickly
10-19	Nothing for you. Return to station.	Return to
10-20	What is your location?	Location
10-21	Call by telephone	Call by telephone
*10-22	Report in person to	Disregard
*10-23	Stand by	Arrived at scene
10-24	Finished with last assignment	Assignment completed
*10-25	Can you contact	Report in person to
10-26	Disregard last information	
*10-27	Go to channel	Drivers license information
*10-28	What is your name?	Vehicle registration information
*10-29	What equipment are you using?	Check records for wanted
10-30	Does not conform to rules and regulations	Illegal use of radio
10-31		Crime in progress
*10-32	I will give you a Radio Club	Man with gun
10-33	Emergency traffic at this station; don't interrupt	Emergency
*10-34	Trouble at this station need help	Riot
10-35	Be alert for	Major crime alert
10-36	Confidential information (correct time, in some areas)	Correct time
*10-37	Wrecker needed at	Investigate suspicious vehicle
10-38	Help—need immediate assistance	
10-39	Proceed to	
10-40	Car to car	
*10-41	Please tune to channel	Beginning tour of duty
*10-42	Traffic accident at	Ending tour of duty
*10-43	Traffic tieup at	Information
*10-44	Message received by all concerned	Request permission to leave patrol for . . . .
*10-45	All units within range please report	Animal carcass in lane at . . . .
10-46		Assist motorist
10-47		Emergency road repairs needed
10-48		Traffic standard needs repairs

10-49		Traffic light out
10-50	Accident	Accident—F, PI, PD
10-51	Wrecker needed	Wrecker needed
10-52	Ambulance needed	Ambulance needed
10-53		Road blocked
10-54		Livestock on highway
10-55		Intoxicated driver
10-56		Intoxicated pedestrian
10-57		Hit-and-run—F, PI, PD
10-58		Direct traffic
10-59	Convoy or escort	Convoy or escort
*10-60	What is next message number?	Squad in vicinity
10-61		Personnel in area
*10-62	Unable to copy, use phone	Reply to message
*10-63	Net direct to	Prepare to make written copy
10-64	Net clear	Message for local delivery
*10-65	Awaiting your next message	New message assignment
10-66		Message cancelation
*10-67	All units comply	Clear to read net message
10-68		Dispatch information
10-69		Message received
10-70	Fire	Fire alarm
10-71	Proceed with transmission in sequence	Advise nature of fire (size, type & contents of building)
10-72		Report progress on fire
*10-73	Speed trap at	Smoke report
10-74		Negative
*10-75	You are causing interference	In contact with
10-76		En route
*10-77	Negative contact	ETA (estimated time of arrival)
10-78		Need assistance
*10-79	Report progress of fire	Notify coroner
10-80		
10-81		
10-82	Reserve hotel room for	Reserve lodging
10-83		
*10-84	What is your telephone number	Are you going to meet . . . . ?
*10-85	My address is	Delayed, due to
10-86	What is your address?	
10-87	Pay checks out	Pick up checks for distribution
10-88	Advise present phone number	Advise phone number to contact
10-89	Radioman needed	
*10-90	I have TVI	Bank alarm
*10-91	Too weak, talk closer to mike	Unnecessary use of radio
10-92	Too loud, talk farther from mike	
10-93	Frequency check	
*10-94	Give a test—with voice or without voice	Drag racing
10-95	Transmitting from base	
*10-96	Transmitting from mobile	Mental subject
10-97	Arrived at	
*10-98	Assignment completed	Prison or jail break
*10-99	Cannot read you	Records indicate wanted or stolen
10-100	Nature calls	
10-200	Police needed at	
10-400	Drop dead	

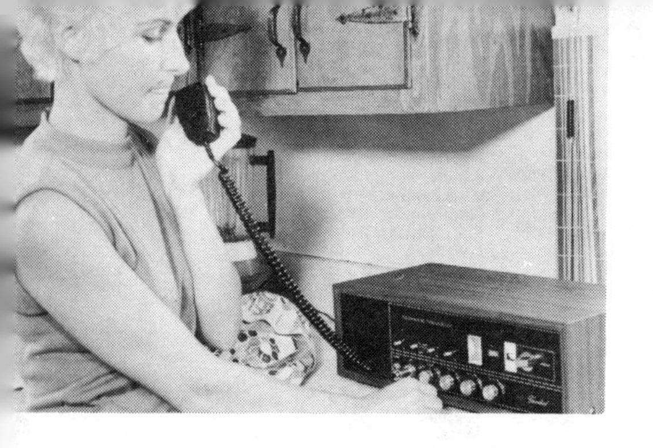

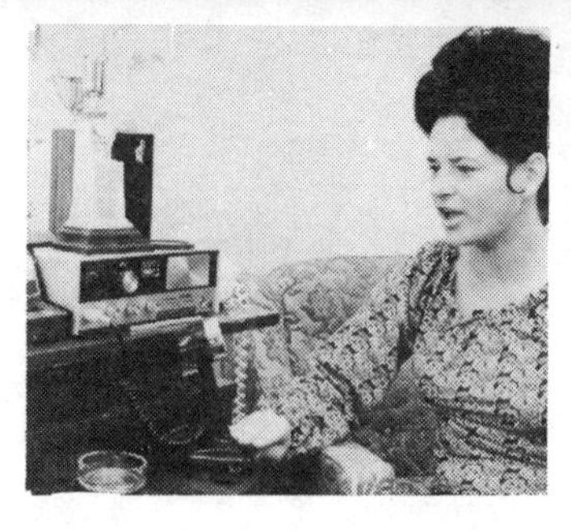

What do you do when you want (or need) to talk on a specific channel? Perhaps you've made arrangements to contact someone at one of your other transceivers at a specific time on a particular channel. In that event, you ask for a "break."

For the first rule of breaking, you try not to "walk on" anybody with your transmission. Wait for a pause between the two conversations, like when they're switching back and forth from one person to the other. Key your mike and say, "Break." Or if you're on channel 14, you might say, "Breaker one-four." Polite CBers will wind up their conversation shortly, and leave you a chance to talk.

If you say, "Short break," or "Break for a short," the other CBers know you need the channel for only a minute or less. They might interrupt their exchanges to let you transact your brief business. A breaker may emphasize that he does not want the channel long by saying "Break for a short, short."

Sometimes you hear selfish, thoughtless CBers who don't want to give up the channel (even when they're violating the five-minute conversation limit imposed by the FCC). That triggers an equally discourteous breaker to shout "break, break, break, break, break" in staccato tones to "step on" the other conversations. Both practices are childish and only immature CBers engage in them.

After you have the channel, here's a typical short break: "This is KEQ3427 mobile unit 1 calling base." Wait a few seconds for a reply, then repeat. If you don't make contact, it's customary to say, "Negative contact. Thanks for the break. I'm on the side."

The phrases "standing by" or "on the side" mean you're listening on that channel in case anyone calls you.

One way to learn CB technical terms is by shopping for "a pair of ears" . . . in other words, for some CB equipment.

Put the word *transceiver* at the top of your list. Maybe you thought you wanted a CB radio; actually, you want a transceiver. Think of the broadcast radio in your car—all you can do with that is listen, or receive. With CB you need to transmit as well as receive. Therefore, a CB unit which does both is a transceiver.

You may need a *mobile* transceiver (for car, truck, boat, or recreational vehicle) or a *base* unit (for home or office). The electricity that powers them distinguishes between the two types. Mobile units operate from a 12-volt dc power source—like the battery in your car. Base transceivers have an electrical cord and plug so they can operate on 117-volt ac house power. A few base-type transceivers can work from either type of voltage.

If you study transceiver specifications (specs), you'll see *output, output power,* or *rf output power.* All refer to watts of rf power the transmitter sends to the antenna. The FCC allows 4 watts. Some CB transmitters put out less; they're not as powerful.

Older specs may list *input* power instead of output, and show 5 watts. Inside the transmitter, this translates to 4 watts of rf output power. So this spec means nothing anymore.

You'll notice the word *sensitivity,* accompanied by a bunch of letters, a number, and a greek mu (μ) beside a capital V. Sensitivity refers to your receiver's ability to amplify incoming CB signals so you can hear weak and distant signals clearly. The μV signifies microvolts. The microvolts number tells you how little signal your receiver can bring in and still make understandable. Therefore, the lower the μV number or sensitivity rating, the better the receiver side of your transceiver.

Selectivity and *adjacent channel rejection* refer to how well your receiver shuts out channels alongside the one you want. You shouldn't hear channel 3 or 5 conversations when you're tuned to channel 4. In spec ratings, the higher the selectivity number (usually expressed as so-many decibels or dB), the better the receiver.

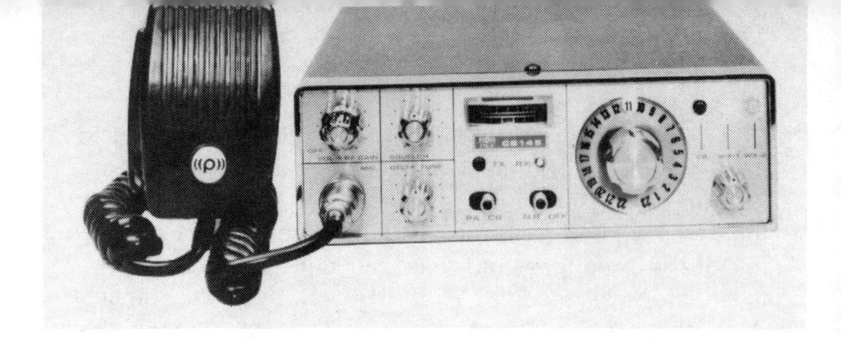

You'll want to know what the features of various CB transceivers refer to.

A few sets have a knob with AM, LSB, and USB settings. AM stands for amplitude modulation, the regular CB operating *mode.* LSB refers to lower sideband, and USB to upper sideband, which are special modes of CB operation. Only *single sideband* transceivers have this mode selector.

To understand operating modes, you need to know a bit about a CB radio signal. Picture an airport with 23 paved runways, side by side, spaced well apart. Regular CB transceivers operating in the a-m mode transmit and receive on one runway (called the *carrier* signal). When you talk, little taxiways appear on both sides of the runway you're talking on, and parallel to them. These taxiways are called *sidebands,* and a-m operation sometimes is called a *double-sideband* mode.

Peculiarly, these taxiways (sidebands) contain the voice information of your CB message. Either one of the taxiways (sidebands) can be sent out ALONE by a special CB transmitter, without the runway (carrier) even going along. Hence, runway 10 could be transmitted by a-m (carrier plus sidebands), by upper sideband (one "taxiway" by itself), or by lower sideband (the other sideband alone).

With the mode knob at AM, single-sideband transceivers can communicate with regular CB outfits. But sideband communication requires a single-sideband (ssb) transceiver at both ends. Ssb transceivers cost more. Your first transceiver will likely be a regular type, unless you have neighbors and friends who have already "gone sideband." (More about ssb on page 31.)

You'll always find a knob labeled *Squelch*. Squelch mainly kills the noises a receiver makes when there's no strong signal coming in. You hear only CB signals that are strong enough to "open" the squelch circuit. Page 83 tells how to set and use your squelch control.

Delta Tune or *Clarifier* helps you match your receiver frequency precisely to another CBer's transmitter. When a transmitter becomes slightly out of tune, reception of it may be slightly ragged. Clear it up by fine-tuning with the Delta knob. Most ssb CB transceivers have Delta controls; but don't think you're necessarily buying a single-sideband transceiver just because it has that feature.

Switches labeled *ANL* and *NB* activate noise-suppression circuits. *Automatic noise limiting* helps reduce sharp noise impulses, like ignition interference. A *noise blanker* cuts them out even better. A few better models have both.

Telephone-type *handsets* have come to CB recently. When the handset rests in its cradle, you hear through a regular speaker. Pick up the handset to talk, and you receive through the earpiece. You transmit by pressing a button in the handset and talking into the mouthpiece.

Some sets sport a *Local/Distant* switch. This switch cuts down receiving distance. If you want to hear only CBers close by switch to Local. If you want the maximum distance from your radio, switch to Distant. It works much like the RF Gain control.

An *S meter* measures relative strength of any signal you receive. When a CBer calls for a radio check, he wants to know where you are and how well you "read" him. Just tell him where your meter registers his signal. If he's a mile or more away and your needle goes past 9—to 20, 30, or 40 dB—he's probably "wearing socks" (operating an illegal linear amplifier—page 32). Most transceiver meters also function as an *RF* or *Power* meter when you're transmitting.

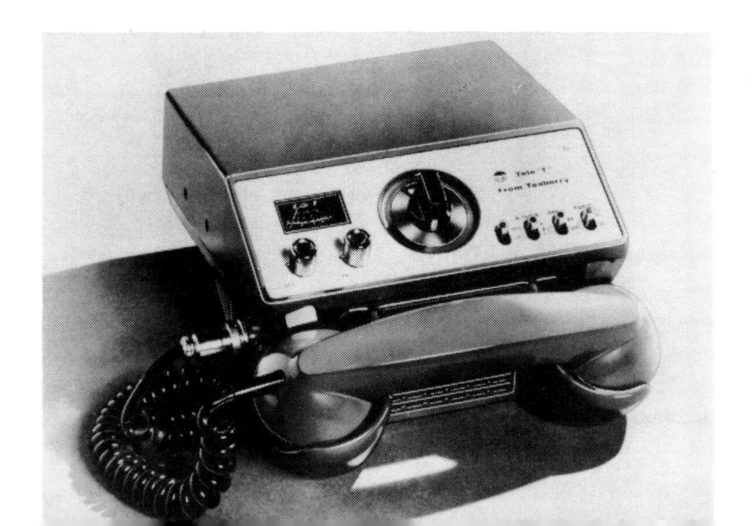

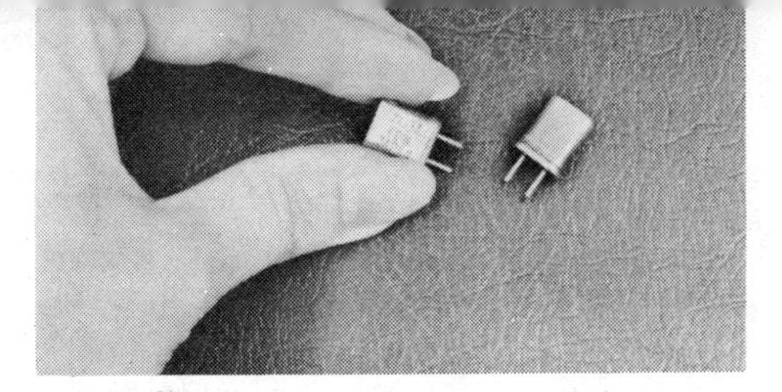

Some CBers, particularly beginners, lack the space or the cash for a 23-channel transceiver. That being the case, consider the small 3-, 5-, 6-, or 12-channel models.

The Federal Communications Commission requires crystal control of all CB transceivers. Crystals are small, can-like devices with plug-in prongs on one end. They lock the set to a particular frequency. A channel-4 transmit crystal at 27.005 MHz, for example, lets the transmitter develop only channel-4 signals. Its companion receive crystal holds the receiver tuned so you hear only channel-4 signals. Hence, it takes two crystals for each CB channel in these small transceivers.

Crystals aren't cheap. If you had to buy two crystals for each of 23 channels, you'd pay out a lot of money. So, today's larger sets are *synthesized.* This term, as it applies to CB, involves mixing the signal from one crystal with the signal from another crystal to form yet a third signal. A synthesized 23-channel set thus can get by with maybe a dozen crystals. And they're already in the set when you buy it.

In 3-, 4-, or 6-channel sets, choose your channels wisely. Always buy a pair of channel-9 crystals for highway emergency. For a mobile unit, get crystals for the trucker channel. Buy a pair for your local "chatter" channel.

Smaller sets that supply 4 watts of transmitter power reach out just as far as large sets. But the few channels do limit your CB activities. Don't despair. A little thought, and maybe some discussion with a local CB expert, can make your limited-channel set almost as useful as a fully synthesized, 23-channel station.

Single-sideband (ssb) Citizens Band radio appears to most CBers as a whole new ball game. It's not necessarily. On preceeding pages we tell about the a-m mode, the rf carrier, and your legal maximum transceiver output (4 watts). Single-sideband goes beyond some of these limitations.

How? Without getting too technical, we can say that ssb transceivers zero in on one sideband of each specific channel, concentrating and boosting power and efficiency there. You develop greater "talking" signal strength, and your CB communications travel farther.

Moreover, you have 42 sidebands to choose from for general conversation—21 upper sidebands, and 21 lower sidebands. That's because there is one on each side of every CB channel. (Reserve channels 9 and 11 for their intended purposes.)

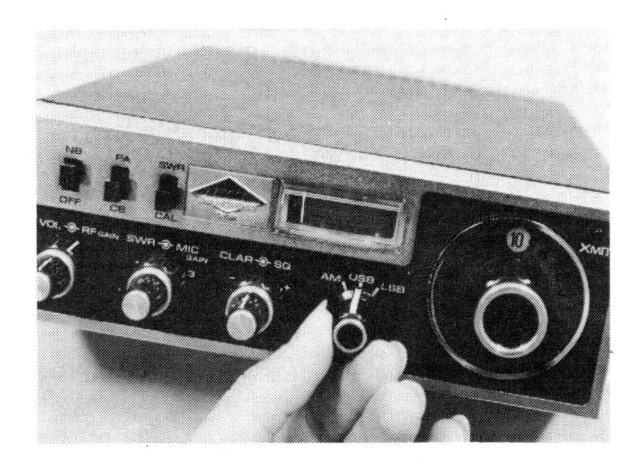

To take advantage of single-sideband, use only ssb transceivers in your personal system. Use upper- or lower-sideband modes during your working hours, when you need peak efficiency. You can join neighborhood gab sessions at night by switching to the a-m mode.

Eventually, single-sideband will distribute traffic more widely on the CB frequencies, so more people can talk at the same time. While you're working the upper sideband of channel 6, someone else can use the channel-6 lower sideband. However, ssb transmissions "step on" a-m talk on the same channel—and vice versa. You cannot carry on a conversation with someone on a-m if you're on ssb. You hear each other only as noise and garbled talk.

Any CBer who "takes a trip" with his "socks, shoes, boots, or moccasins on," or who "has a tiger in his tank," operates illegally. He's using a *linear amplifier.* A linear boosts output wattage far beyond the maximum 4 watts authorized by the FCC.

With this added power, the impact of the CB signal is greater, although it doesn't travel all that much farther. CB waves follow mainly a line-of-sight path. As long as your signal isn't blocked by tall buildings, or you're not surrounded by hills or mountains, you can reach out quite nicely for your 4 watts. You may sound a bit weak around the fringes of your coverage area, but anyone within 5-10 miles can hear you well enough for easy communication.

All a linear amp does is amplify the rf signal that carries your voice modulation. Sometimes that blots out other channels (covers them with the amplified transmission), rendering conversation on them inaudible.

Linear amps provide insecure CBers a kind of ego-trip in one-upmanship. It's both illegal and grossly discourteous to use one. If the FCC discovers you using, buying, selling, or even storing a linear amplifier, you're in for trouble (if you call losing your CB gear and license, and incurring a possible fine and prison term big trouble).

So if you want to walk on the safe side, go barefoot (your unadorned 4 watts). You'll find plenty of other CBers to talk to in the same unshod state. If you absolutely need more distance, put up a taller or better antenna. It's cheaper all the way around.

When you start looking for an antenna, advertising and salesmen hurl words like *omnidirectional, beam, gain, ground plane, dipole*—and others. The most important factor for performance is antenna height. Whether the antenna goes on your car or beside your home, the higher its tip rises above the earth's surface, the farther your station sends and receives. In Chapters 4, 6, and 7, we show you how to install various types of antennas. But for now, let's delve into some of the technical words that surround antennas.

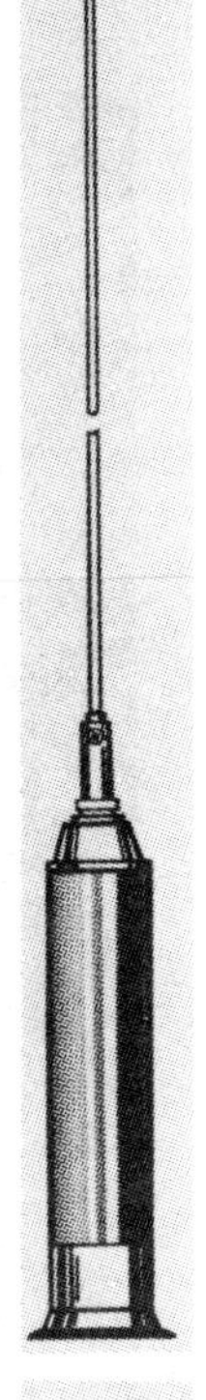

Most CBers begin with a mobile station, adding a home base later—when they've discovered CB's merits. So we'll discuss mobile antennas first.

The ideal length for a mobile antenna is 102 inches. A 102-inch *whip,* as the steel rod is called, fits the quarter-wavelength size that is efficient for transmitting and receiving CB frequencies. But 8 1/2 feet of antenna towering over a car or truck is impractical.

Antenna engineers have found they can "load" a shorter antenna by winding a coil and adding it in series with a short vertical antenna rod. This produces the quarter-wavelength needed, but electronically instead of physically. Today, you'll see mobile antennas with rods as short as 18 inches. All have a loading coil somewhere.

Coils in *base-loaded* antennas, for example, are at the bottom of the antenna rod. In *center-loaded* models, the coil is situated along the midsection of the rod, but not necessarily in the exact center. Some manufacturers wrap the entire length of the vertical element with the loading wire. Such loading is *helical.*

Top-loading is something different. A coil at bottom or center extends the rod's electronic wavelength beyond the quarter-wave size. Then a movable metal ball or array at the tip of the rod "trims" the wavelength back to precise wavelength. It's done so a technician can tune the antenna for peak efficiency.

The majority of mobile antennas today come packaged with mounting kits. With no-hole quick-mount types (Chapter 6) you have everything you need, including the coaxial lead-in cable.

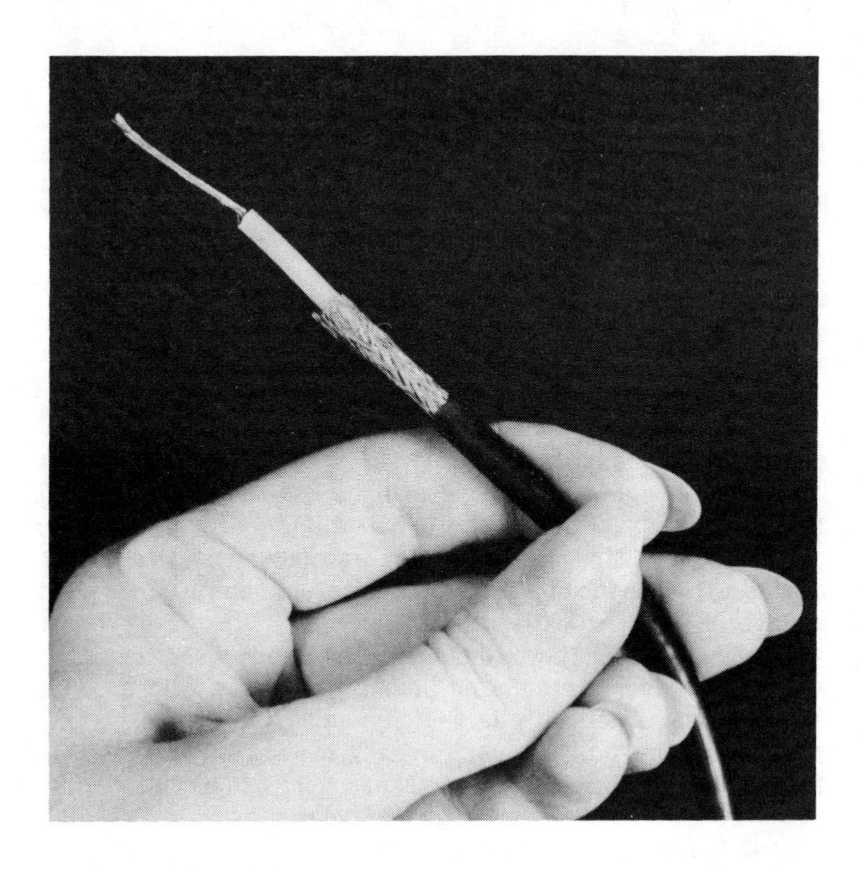

The "coax" cable that connects your antenna to your transceiver is not just any insulated wire, but rather two wires in one. Polystyrene insulation surrounds a center wire, or core conductor, made of copper. A second conductor, a metallic braid called the *shield,* fits outside the center insulation. The polystyrene separates the center conductor and the shield, keeping them a fixed distance apart. That prevents their shorting out electrically. The spacing also maintains a characteristic called *impedance,* which matches the CB transceiver's rf circuits to the CB antenna.

The black vinyl outer cover insulates and protects the shield.

RG-58/U coaxial cable presents 50 ohms impedance to your antenna and your receiver. Never replace the CB coax with any other type of wire. You'll mess up something if you do.

You might become a bit discombobulated trying to select the right home-station antenna. Two considerations help you decide what to buy: the space you have available, and the money. Limited both immediately suggest an omnidirectional *ground-plane* antenna.

Omnidirectional antennas send and receive CB signals equally in all directions. Placed as high as the FCC allows (60 feet from the ground to the tip of the antenna), an ordinary ground-plane antenna is efficient enough for even the best CB operation.

But, if you've saved up some money and want something a bit more, consider an omnidirectional antenna with *gain.* That term refers to the ability of certain antennas to in effect multiply the 4 watts of signal put out by your transceiver. Your CB station radiates more power *through* your antenna.

For example, one antenna spec sheet claims a 4-dB gain. Four watts from your transceiver increases to about 13 watts of effective radiated power (erp) with that antenna.

An interesting phenomenon occurs when you raise the antenna height or put up a higher-gain antenna. Every improvement your antenna system makes for transmitting shows up also as an improvement for receiving.

If you want still more radiated power, go to a *directional* or *beam* antenna. It looks impressive on your roof, and does improve reception *and* transmission. However, because a beam antenna concentrates reception sensitivity and radiated power in one direction, you lose range in other directions.

You lose something else with a beam. The FCC permits an omnidirectional antenna to go as high as 60 feet, if there's no airport nearby. However, a beam antenna must reach no more than 20 feet above the top of your house.

Most of the terms written into antenna spec sheets take an engineer to understand. Yet, you need to know a little bit of what those specs are talking about. Otherwise, you can't cut through the advertising hype to the nuts-and-bolts. So, herewith we bring explanations of some important antenna terms.

Angle of radiation: Upward direction radio waves leave the antenna array. The lower the angle, the better the antenna, because not so much power is wasted "off into space."

Dipole: Two-element antenna. In CB antennas, which are vertically oriented, the top element usually is active (see Element). Dipole antennas are usually cut to half-wave dimensions (also see Wavelength).

Element, active: Also called *radiator* and *driven element.* Antenna rods. The coaxial feedline from the transmitter connects to it, and it is the rod that radiates the actual electromagnetic radio-wave energy into the surrounding air.

Elements, passive: Also called *parasitic elements. Reflectors* are rods behind the driven element; they reinforce radiation in the opposite direction. *Director* rods, in front of the driven element, reinforce in that direction. *Radials* of a ground plane are passive elements, too (see Ground Plane).

Front-to-back ratio: An indicator of how directional a particular beam antenna is. The greater the f/b ratio, the more efficient the beam.

Gain: A measure of how much the antenna raises the effective power radiated, when compared to a reference (either a standard ground plane or what's known as an *isotropic* source—an engineer's standard). Usually expressed in decibels (dB). Can be converted roughly to a factor for *effective radiated power* (erp). For example, 3 dB of antenna gain multiplies transmitter power by 2.7; 4 dB multiplies power by 3.3. For a gain of 5 dB, multiply output power by 4.3 to estimate erp; for 6 dB, transmit power times 5.3; for 8 dB, times 8.4; and so on.

Gamma match: A short bar between antenna-lead connection point and a precalculated spot on the driven element. Purpose is to match inpedance of the antenna radiator precisely to that of the coaxial cable that runs to the transceiver. Without this match, vswr (page 29) would be high and the antenna system would waste transmitter power instead of radiating it efficiently.

Ground plane: An antenna characterized by one vertical rod (the radiator) and four horizontal radials (which are the ground plane). Radials act as a reflector for the radiator, affecting its angle of radiation and its gain (if any). Simple ground-plane antennas, if efficient, radiate about 75 percent of the power the transmitter puts out.

Wavelength, halfwave, quarterwave: Usually refers to the electronic "size" of the radiating element of an antenna. Mobile whips are designed to be quarterwave. Base antennas can be quarterwave (ground planes usually are) or halfwave (as most "gain" antennas are, whether omnidirectional or beam-type). Wavelength, for antenna purposes, can be calculated by dividing 935 by operating frequency in megahertz. Hence, for CB at 27 MHz, a halfwave antenna measures about 17.3 feet.

Wing-span: Space needed for mounting a CB antenna. For omnidirectionals, determined by length of radials. For beams, by length of the boom. With beams, also allow room for rotating.

Whether you put up a huge antenna or the cheapest one, occasionally you'll pick up someone from halfway across the country. It's really a thrill to hear someone "lookin' " from Albuquerque when you live in Buffalo.

You'll be tempted to "come back" on the call, just to see if you can reach that far. But don't. Under section 95.83 (16)b of the FCC Rules and Regulations you must not communicate with anyone beyond a distance of 150 miles.

Such extra-long-distance transmissions on CB frequencies are called *skip*. Freakish atmospheric conditions, often sunspots or sunstorms, alter electrical balance in the ionosphere (that layer of atmosphere which lies from 50 to about 250 miles above the earth's surface). When the ionosphere becomes electrically charged just so, CB signals ricochet off it like a bullet. Some of your transmitter signal bounces back down to earth some 400 to 600 miles away from your transmitter. Under some conditions, your signal bounces repeatedly; CBers a couple thousand miles away may hear your voice. And you can hear their bouncing signals.

The phenomenon, of course, no one can stop or control. But the FCC does forbid using skip for CB communication.

Nevertheless, over Channel 19 some frosty March evening, Honeysuckle in Tuscaloosa may stumble into a signal from Branch Rickrock on California's northern coast. The chance meeting may warm both their hearts. But the romance will be short-lived, because skip is unreliable. You'd do better to stick with local CBing. If you want to talk long distances, legally, become a ham (radio operator, that is).

Chapter 3

Getting Your CB License

Not long ago, a class-D Citizens Band license cost $20. Today, it's only $4. And the license is good for the same five-year duration.

Also, in those days, filling out the application form was a major chore. You had a full sheet of instructions to read before you even began. And there was a work sheet so you could work out the right answers and then transfer them to the application form without unsightly erasures or changes to confuse the FCC office and put a hold on your application.

But now Form 505 has changed to a simple, one-sheet deal. You'll find no tricky questions, none of those confusing legal words that are so easily misunderstood. In years past, applications were rejected frequently because a puzzled applicant filled a blank incorrectly.

Our next few pages explain the new Form 505 in detail. Even though filling it in may seem like child's play, it's important to do it right. If you'll follow the procedure we outline, your CB license should arrive within four to six weeks.

Don't put off applying for your CB license. With more than 100,000 others sending in Form 505's some months, turnaround time may grow longer. And you cannot operate your CB station or mobile legally without a license.

You can pick up FCC Form 505 (the license application) at almost any CB shop. If your local CB dealers have none, write your nearest FCC Field Office. You will receive, in a few days, a Form 505 blank, along with bulletins and information regarding the Citizens Radio Service. There's no charge for either the application form or the bulletins.

FCC FIELD OFFICES

439 U.S. Courthouse & Customhouse
113 St. Joseph Street
Mobile AL 36602
205-433-3581, Ext. 209

U.S. Post Office Building
Room G63
4th and G Street
P.O. Box 644
Anchorage AK 99510
907-272-1822

U.S. Courthouse
Room 1754
312 North Spring Street
Los Angeles CA 90012
213-688-3276

Fox Theatre Building
1245 Seventh Avenue
San Diego CA 92101

300 South Ferry Street
Terminal Island
San Pedro CA 90731
213-831-9281

323A Customhouse
555 Battery Street
San Francisco CA 94111
415-556-7700

504 New Customhouse
19th St. between Cal. & Stout Sts.
Denver CO 80202
303-837-4054

Room 216
1919 M Street, N.W.
Washington DC 20554
202-632-7000

919 Federal Building
51 S.W. First Avenue
Miami FL 33130
305-350-5541

738 Federal Building
500 Zack Street
Tampa FL 33606
813-228-7711, Ext. 233

1602 Gas Light Tower
235 Peachtree Street, N.E.
Atlanta GA 30303
404-526-6381

238 Federal Ofc. Bldg. & Courthouse
Bull and State Streets
P.O. Box 8004
Savannah GA 31402
912-232-4321, Ext. 320

502 Federal Building
P.O. Box 1021
Honolulu HI 96808
546-5640

37th Floor - Federal Bldg.
219 South Dearborn Street
Chicago IL 60604
312-353-5386

FCC FIELD OFFICES GUIDE

829 Federal Building South
600 South Street
New Orleans LA 70130
504-527-2094

George M. Fallon Federal Bldg.
Room 819
31 Hopkins Plaza
Baltimore MD 21201
301-962-2727

1600 Customhouse
India & State Streets
Boston MA 02109
617-223-6608

1054 Federal Building
Washington Blvd. & LaFayette Street
Detroit MI 48226
313-226-6077

691 Federal Building
4th & Robert Streets
St. Paul MN 55101
612-725-7819

1703 Federal Building
601 East 12th Street
Kansas City MO 64106
816-374-5526

905 Federal Building
111 W. Huron St. at Delaware Ave.
Buffalo NY 14202
716-842-3216

748 Federal Building
641 Washington Street
New York NY 10014
212-620-5745

314 Multnomah Building
319 S.W. Pine Street
Portland OR 97204
503-221-3097

1005 U.S. Customhouse
2nd & Chestnut Streets
Philadelphia PA 19106
215-597-4410

U.S. Post Office & Courthouse
Room 322 - 323
P.O. Box 2987
San Juan PR 00903
809-722-4562

323 Federal Building
300 Willow Street
Beaumont TX 77701
713-838-0271, Ext. 317

Federal Building - U.S. Courthouse
Room 13E7
1100 Commerce Street
Dallas TX 75202
214-749-3243

5636 Federal Building
515 Rusk Avenue
Houston TX 77002
713-226-4306

Military Circle
870 North Military Highway
Norfolk VA 23502
703-420-5100

8012 Federal Office Building
909 First Avenue
Seattle WA 98104
206-442-7653

An application form frequently comes in the box with new CB transceivers too. But if you get your application this way or from a CB shop, you miss out on the bulletin goodies from the FCC.

One you have Form 505 in hand, read it carefully before you begin. It looks simple, and it is. But to avoid trouble with Federal authorities, you must agree to a few specific requirements.

Item 10 (15 on some forms), labeled *Certification,* contains the first chore to deal with. Look at the second point. You swear that you own, or have ordered from the Government Printing Office, an up-to-date copy of Part 95 of the FCC Rules and Regulations. What you actually must buy is a copy of Volume VI of the FCC Rules and Regulations; it includes Part 95, detailing class-D Citizens Band operation.

Also note that this requirement also applies at license renewal time. And any Part 95 that is five years old is out-of-date and misleading.

We've talked to CBers who haven't seen a copy of the Rules for 18 years. You'll meet some too. Don't follow their poor example. Neglecting to have a copy of current FCC regulations is a double-edged sword. Not only do you remain unaware of new relaxed laws, or of new legal antenna heights, but you are also ignorant of the new violations you may be inadvertently committing. Your ignorance won't save you a nickel if you're caught in violation by the FCC. So be smart. Stay informed.

Fill in the order blank at the bottom of Form 505, if your version has it. Attach a check or money order. As we write this, the price for Volume VI (including Part 95) is $5.35. It does change from time to time, so don't go by outdated forms. You'll note that the order form does say "subscription." This initial fee includes one year of updating bulletins.

When you've completed the order form or your order letter, put in your check or money order; mail them to: Superintendent of Documents, Government Printing Office, Washington DC 20402. DO NOT send your request for Rules and Regulations to the FCC. They don't want it, and you may never see Part 95 if you do.

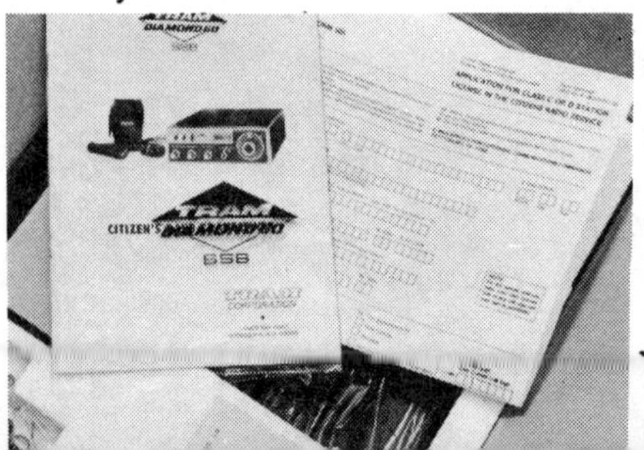

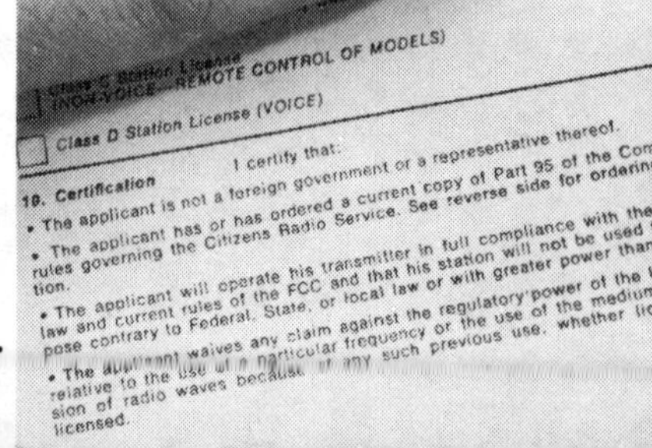

REMOTE CONTROL OF MODELS)

Class D Station License (VOICE)

10. Certification — I certify that:

- The applicant is not a foreign government or a representative thereof.
- The applicant has or has ordered a current copy of Part 95 of the Com[...] rules governing the Citizens Radio Service. See reverse side for ordering [...]tion.
- The applicant will operate his transmitter in full compliance with the [...] law and current rules of the FCC and that his station will not be used fo[...] pose contrary to Federal, State, or local law or with greater power than [...]
- The applicant waives any claim against the regulatory power of the U[...] relative to the use of a particular frequency or the use of the medium [...]sion of radio waves because of any such previous use, whether lice[...] licensed.

When Volume VI arrives, read Part 95 thoroughly. Some of it may not make much sense to you at first, particularly if you haven't received your license (and consequently started transmitting). You'll understand more when you work the CB channels a bit. (You'll also discover how few CBers really know the newest laws.)

Keep Part 95 handy, not only for information, but just in case an FCC field man decides to drop in at your station. Sections 95.103 and 95.105 apprise you that an FCC official can drop in to inspect your station anytime (just as they do at commercial television and radio stations). Your license should be posted by your base station.

If your only transmitter is mobile, you keep your license at the address listed on your license and you tag the CB radio in your car. The tag lists station call sign, and the name and address of the licensee (you). The FCC will send Form 452C for that purpose, if you ask (page 110).

The more you become involved with CB, the more the Rules and Regulations will mean to you. These Rules set out guidelines for CB operation so a maximum number of people can enjoy the limited number of frequencies available. They give us all a chance to get our two-cents-worth in. Including you.

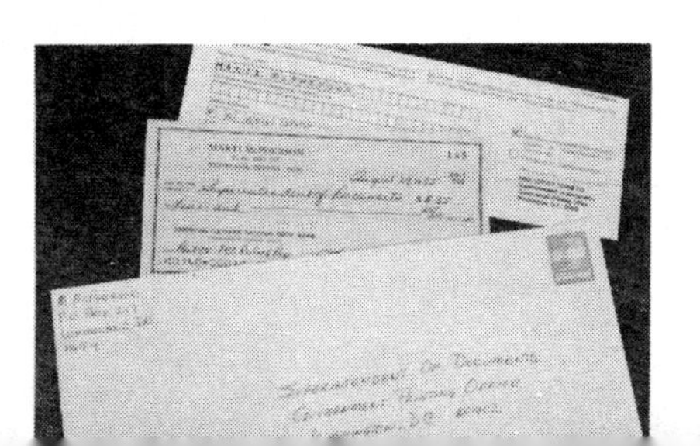

To fill out the newest (August 1975) version of Form 505, either print the information with a pen or use a typewriter. Use capital letters. Leave blank blocks where spaces would appear if written regularly. Here are some other, specific hints. (Note that the numbering differs on some forms.)

If you do not put a space between the C and the Q normally, don't do it on the form. If you do, leave a blank space after the C. (The same applies to a Post Office box or RFD number in *Item 4.* POBOX jammed together is poorly legible.)

Item 2: You must be at least 18 years old to get a Class-D Citizens Band license. This rule may be modified soon, lowering the minimum age to 16; but so far this move has not been made by the FCC. Incidentally, don't forget that it's your date AND YEAR of birth, not today's date or this year, that belong in item 2.

If you're applying as an association, corporation, or government entity, you don't fill in a date of birth.

Item 3: If you're applying as a business, print the full business name here.

Item 4: Fill in your mailing address (refer to item 1).

Item 5: If your mailing address is either a Post Office box or an RFD number, you must show the location of your principal (base) station, or where your station license will be kept if your only transmitter is in a car or truck. If you do not have a street address, give directions from major roads. The description might read something like: 2 1/4 miles east of County Road 450 on Old Mill Pike. You can also use longitude/latitude coordinates. Someone at the county courthouse or land office can tell you what they are at your "home 20."

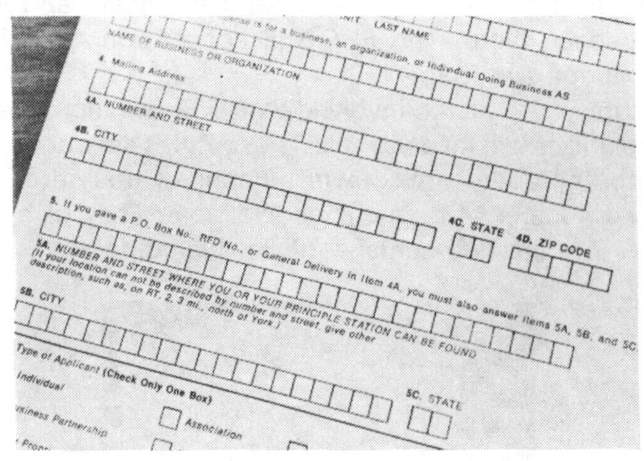

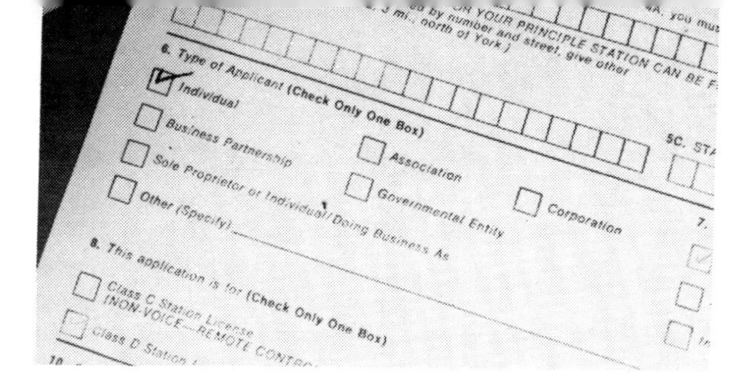
6. Type of Applicant (Check Only One Box)
Individual
Business Partnership
Association
Sole Proprietor or Individual/Doing Business As
Governmental Entity
Corporation
Other (Specify)
8. This application is for (Check Only One Box)
Class C Station License
(NON-VOICE—REMOTE

Item 6: As you can see, individuals are not the only ones applying for CB licenses. Companies and associations use them too.

If you are applying as an individual, so indicate in the square. Only one person in any household needs a license. Therefore, once yours arrives, you can authorize anyone living under your roof to use your call sign and equipment. Your license covers you and your station. It is not related to your equipment. You may have as many operators and as many different kinds of or changes in equipment as you want (see *Item 9,* too). But remember that you are the station owner of record, and are responsible for actions of all your station operators and all the equipment used under your license.

If you are sole proprietor of a business, or if your license is in the name of a corporation or partnership, here's something to remember: Your employees can use your station and equipment legally only when conducting your company business. They cannot use your CB call sign for personal communications, such as to call their own homes. For that they must have their own license.

Item 7: You either want a new license or a renewal, or you want to increase the number of transmitters indicated on your present license. If you're applying for either of the latter two, don't forget to list your current call sign. Mark the appropriate blocks and proceed to *Item 8.*

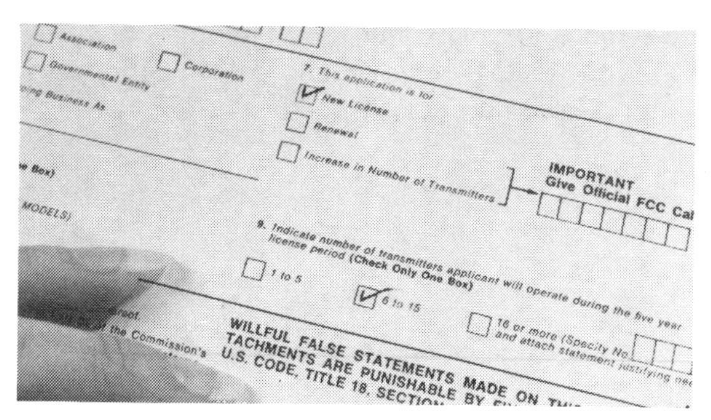
Association
Governmental Entity
Corporation
7. This application is for
New License
Renewal
Increase in Number of Transmitters
IMPORTANT
Give Official FCC
9. Indicate number of transmitters applicant will operate during the five year license period (Check Only One Box)
1 to 5
6 to 15
16 or more (Specify No.
and attach statement
WILLFUL FALSE STATEMENTS MADE ON
TACHMENTS ARE PUNISHABLE
U.S. CODE, TITLE 18,

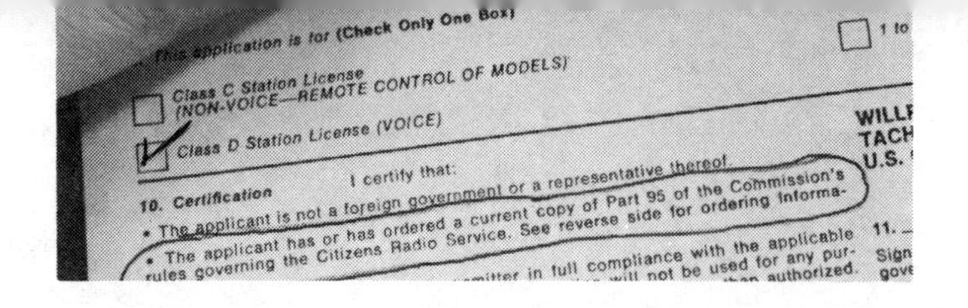

Item 8: For the type of Citizens Band radio we're talking about in this book, and the kind you're becoming familiar with, which has Smokey reports and neighborhood "ratchet-jaw" sessions, you need a class-D station license for voice communications. So put your mark in the class-D square.

Class-C two-way radio is for nonvoice remote control, usually of model boats or planes. If you're a model airplane or boat enthusiast, and a CBer as well, you'll need a separate license for each type of operation. That means two different applications, and two fees.

Item 9: You must indicate the maximum of transmitters (transceivers, actually) you want to use at any time during the term of your license (the next five years).

An individual probably doesn't need more than five transceivers. That would cover a base station, two mobile stations, and leave two for boat, motorhome, or CB walkie-talkies (page 122).

If you're a farmer or rancher, however, you might need between 6 and 15 units—particularly if you have a lot of land. You'd need a base station, a unit in each barn, perhaps one set for PA (public address) in the barnyard area, one for each tractor, combine, and other large farm implement. And don't forget cars and trucks. You might even have a small plane for crop-dusting or range inspection.

If you need 16 or more transceivers, you must explain why. It's conceivable that a large ranch might need two or three dozen. A company located in a city and needing more than 16 transmitters might be better off with Business Radio. It costs a little more than CB, but fewer people occupy BRS frequencies. If you need two-way radio a lot of the day, class-D CB might not allow you enough access.

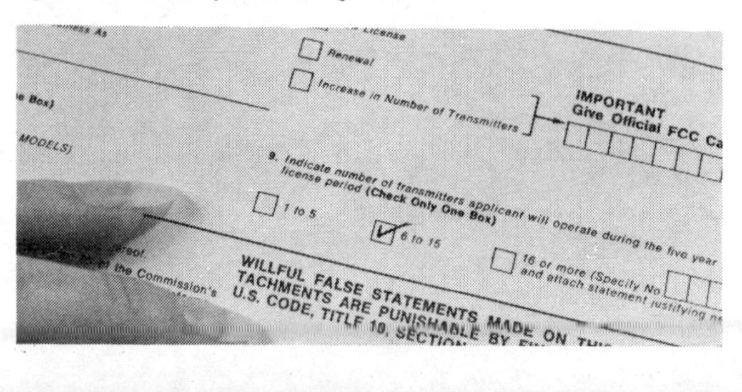

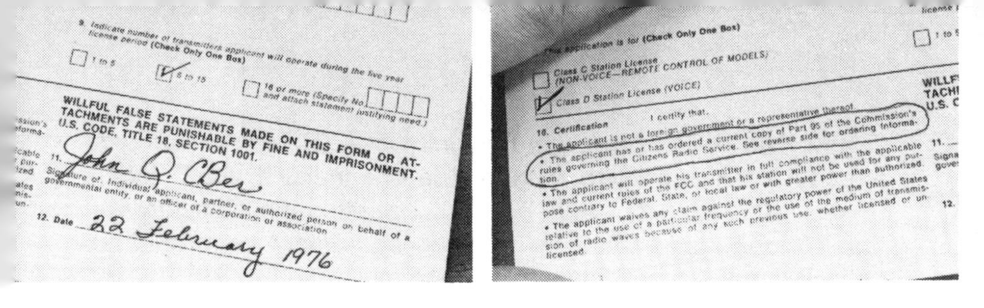

9. Indicate number of transmitters applicant will operate during the five year license period (Check Only One Box)

☐ 1 to 5 ☑ 6 to 15 ☐ 16 or more (Specify No. and attach statement justifying need.)

WILLFUL FALSE STATEMENTS MADE ON THIS FORM OR ATTACHMENTS ARE PUNISHABLE BY FINE AND IMPRISONMENT. U.S. CODE, TITLE 18, SECTION 1001.

11. John Q. CBer

Signature of individual applicant, partner, or authorized person on behalf of a governmental entity, or an officer of a corporation or association

12. Date 22 February 1976

This application is for (Check Only One Box)

☐ Class C Station License (NON-VOICE—REMOTE CONTROL OF MODELS)

☑ Class D Station License (VOICE)

10. Certification I certify that:

• The applicant is not a foreign government or a representative thereof.

• The applicant has or has ordered a current copy of Part 95 of the Commission's rules governing the Citizens Radio Service. See reverse side for ordering information.

• The applicant will operate his transmitter in full compliance with the applicable law and current rules of the FCC and that his station will not be used for any purpose contrary to Federal, State, or local law or with greater power than authorized.

• The applicant waives any claim against the regulatory power of the United States relative to the use of a [illegible] frequency or the use of the medium of transmission of radio waves because of any such previous use, whether licensed or [illegible] licensed.

Before you sign, seal, and deliver your application to the mail box, read again what your signature certifies.

First off, you swear that you are neither a foreign government nor an agent of one. Next, you affirm that you have, or have sent for, a copy of (Volume VI) Part 95 of the FCC Rules and Regulations. If you haven't, do it now before you forget. (If you send Form 505 to the FCC with that order blank at the bottom still attached, they'll immediately suspect that you don't have, nor have you requested, a current copy of the rules. It may slow up your license approval.)

You also agree to operate your station in accordance with the law. This includes having Part 95 at your station. It also means you will keep your equipment (transmitters and antenna) operating within the Rules and Regulations. So . . . no linear amplifiers to boost your signal.

You say you won't sue the FCC or the United States government if you cannot get access to the CB frequencies because someone else is already using them. Your license doesn't guarantee you the channels; it only permits you to use them as and when you can. You waive any right to sue the government over some unlicensed operator undermining your legal right to operate. (But, you can complain to the FCC; and if they catch the violator, he's in for a stiff penalty: a fine as high as $10,000 and up to a year imprisonment.)

Finally, you acknowledge that if you have lied on your application, and are found in violation, you can be fined or imprisoned.

Knowing all this, sign and date your application. Make out your check or money order (don't send cash) to the Federal Communications Commission, and mail it to Box 1010, Gettysburg PA 17325. Never mail a CB application to a field office or to Washington.

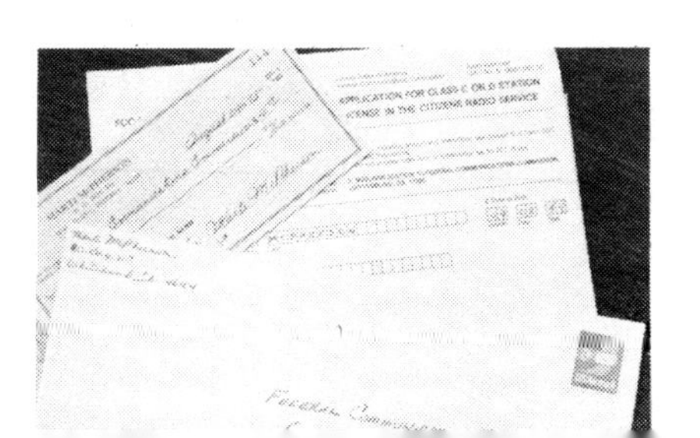

Now you wait. You cannot legally use CB equipment you already have, except to monitor the channels. If you'll recall, there's a notation on your Form 505 application advising you that to operate (transmit) on your equipment is prohibited until you have your license in hand and posted. Unlicensed operation can earn you a fine up to $10,000 and up to a year in prison. So don't do it.

Take advantage of your waiting period to find out what's happening on the various channels. Listen to what other CBers are saying. You'll hear new words and expressions. It's often quite a comedy show. CBers seem to enjoy ribbing each other, all in good fun (usually). Tape your microphone over to remind you that you cannot legally talk yet, or just don't connect the mike. That helps allay the temptation to join in.

As you monitor the channels, refer to Chapter 2 and 5 for some insight into what's being said. Then when your license arrives, you can move right onto the channels like an old pro, even though it's your very first transmission.

Chapter 4

Hooking Up Your Home Station

A lot of CBers begin with a mobile unit in the car. But a home base station comprises the heart of any family CB communications system. If you're the person stuck at home, CB keeps you in touch with neighborhood news. A fire or accident nearby, you know about it shortly. If there's a terrific sale at the shopping center, someone will tell you of it. You can also hear about any severe storms headed your way, in time to batten down the hatches.

As for keeping in touch with the rest of the family, as they pursue their activities around town, we would be redundant to say it's great. Any family member with a mobile or portable transceiver can keep in touch as long as they are not too far away. Teenage drivers should have a CB rig in the car; it can lift a lot of worry from your shoulders. CB drivers can let the folks at home know when road conditions slow them down. You feel less anxiety if they know traffic is holding your guy or gal back, rather than an accident. Only on CB can you get this news easily.

If you're still not convinced about having a home station, go back and read Chapter 1 before you continue with this chapter.

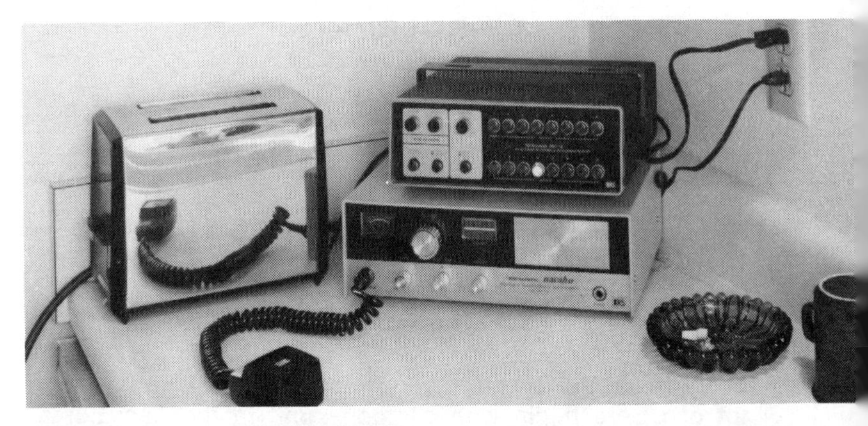

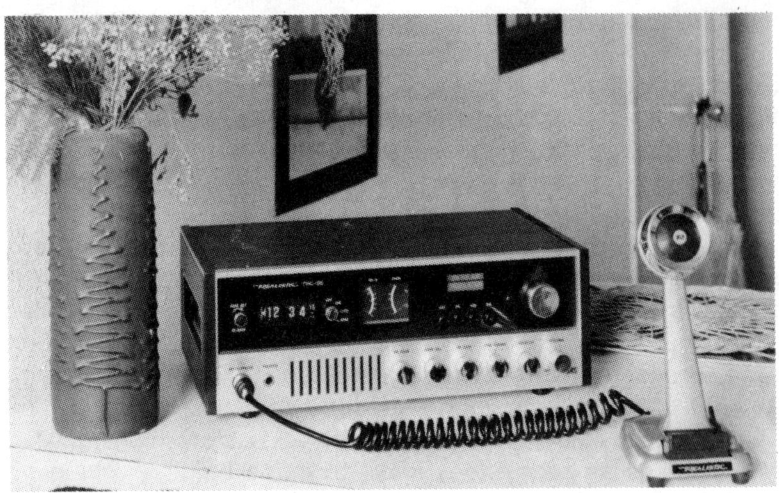

Give careful thought to where you hook up your home station. Just any old place won't do. The workshop down in the basement may seem ideal, at first; but is it near enough to the antenna (page 56)? You should install your base transceiver as close to its antenna as possible, to keep the antenna cable short. More than 60 feet from your transceiver to antenna calls for costly RG-8/U antenna cable rather than inexpensive RG-58/U. Otherwise, much power—and thus range—is lost in the coax cable.

Who will use the base, and when? Any homemaker who spends a lot of time in the kitchen finds that location handy. But let's face it, today's woman spends less time in the kitchen than women did a few years ago. A base in the kitchen has other drawbacks too. It takes up scarce counter space. When you've got your hands in flour, you'll mess up the mike if you have to key it in a hurry—unless it's a tabletop-style mike. If another family member wants to yak on CB while you're preparing a meal—well, there's not always that much room.

Don't set the base too near the stove. You know how everything in the cooking area acquires a coating of yukky grease in a few weeks. That kind of guck wouldn't do much for your base. In fact, the sticky film could cause keying problems in your mike. There must be a better location for your base.

If you decide against the kitchen, what about the family room? Does it offer easy access to the antenna? Everyone can get to the family room easily, and it's not so much in the way of functional household activities. It's an esthetic location for the family base station.

For late-night CBers, what about the bedroom? You can relax and unwind with your CB before dropping off to sleep. But this is not for an early riser in the same room with a night-owl CBer.

One final consideration—and for many families it should be the first. We like to think of CBing as a family activity. But the truth is, some people despise the constant chatter that monitoring a CB channel brings into a room. It constitutes an invasion of their privacy and they grow edgy and short-tempered. That being the case, find yourself a secluded little nook where you can enjoy your CB activities without intruding on other members of your household. Consider headphones to cut the intrusion even further.

Having decided *where* to put your base station, think now about *what* equipment you want to include.

Perhaps for you, a small transceiver on a bedroom nighttable is enough. However, we don't advise using a 3- or 6-channel CB unit for a base radio. You should have the full 23 channels available if you work a base at all.

Or, you might go the elite route, with a fancy CB base station, police scanner, shortwave radio. Maybe a single-sideband transceiver suits you best. Some base outfits have a digital clock with a buzzer to remind you when your time's up (the FCC 5-minute ruling—page 75).

As for power, remember the FCC limits you to a standard 4 watts output (or 12 watts p.e.p. on single sideband). Some unfortunates still violate this ruling by using a linear amplifier, but they're not legitimate CBers. They abuse the privilege of this inexpensive means of communication, and give CB radio a bad name. Linear users are selfish. Their linears trample all over legal transmissions.

You may even feel that, just because some ratchet-jaw down the street "wears socks," you have to also in order to compete. Avoid it. You'll be breaking the law too. One bright night, the FCC will sweep through and point a df (direction finder) at the linear users. You don't want to be one of those paying a fine. Stay clean and barefoot. Omit the linear amplifier from your CB base station shopping list.

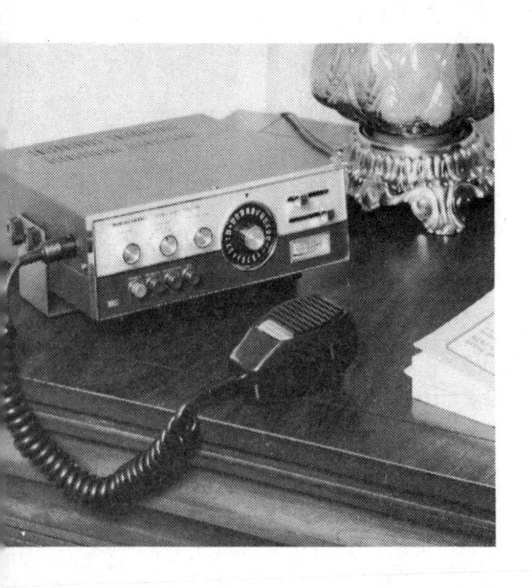

One question keeps coming up in CB discussions: "How far?" Most of the answer lies, as you know by now, in your antenna.

Not everyone can manage a good outdoor CB antenna. Perhaps you live in an apartment or townhouse where outdoor antennas are prohibited. You can buy a small antenna that attaches directly to the back of your transceiver. Cost is under $10.

You won't get a great match to your transceiver. But you can hear some CBers up to 15 miles away if no obstructions block their signals. You won't hear others who have a small antenna like yours. You will hear CBers with tall outdoor antennas. Transmitting, you'll not reach very far. The high vswr of this "monopole" antenna precludes any real transmit efficiency.

Some apartment dwellers who live above first floor level hook up mobile whips outside one of the windows. Efficiency is mediocre; mobile whips need the metal car body for proper operation.

With a tall outdoor omnidirectional antenna, you might possibly send and receive for 20 miles or more under ideal conditions. And, as page 35 points out, an all-direction antenna does not give the range of a beam antenna.

Evaluate who you need to talk with most frequently. How far away will they be? In one direction or many? Having weighed the demands on your transceiver and antenna, you're ready to go shopping for equipment.

There are several places to buy CB equipment. You can browse mail-order catalogs to gain a little insight on specifications. Pictures show what various models look like. Esthetics of a base unit may be important to you. If you live in a very modern home, silver transceiver face, knobs, and dials enhance your decor. Among traditional furnishings, a base with a wood cabinet fits more appropriately. If the base goes in a bedroom, think about a transceiver with a clock in it. Catalogs show what's available.

But you can't tell from a catalog how a set sounds or feels. For this you need a CB specialty shop. You'll find a generous stock of CB equipment at some shops. Others carry only a few brands or models; CB units today sell as fast as they come in. Anyway, try a store where you can get your hands on some equipment. Study the features of each model. Get the feel of their controls before you decide. Then, if you want to order an identical model from a catalog, do so.

BUT, don't overlook the need for service. A catalog can't supply advice or help. Usually, a good CB shop will install or help you install a unit you buy from them, at some minimal charge. Then, if anything goes wrong, you have someone to look to for a cure.

You may find a bargain transceiver at a discount store. It may even be some popular brand and model. But a discounter probably can't install or service your outfit. Repairs, even in warranty, must be shipped away to some factory or service center. You pay the cost of shipping, even if repairs are free. So take that into consideration as you eye the "wholesale" price.

Another (and least recommended) place to pick up CB equipment is from another CBer. True, he may sell his smaller unit so he can buy a bigger one. But you can't be sure he's not selling it because it doesn't work well. If you want to buy some item of secondhand equipment—and you do find an occasional good deal that way—take the unit to an FCC licensed CB technician.

Well, you got home with your new (or used) CB outfit. Unpack it carefully. You'll have the base transceiver, a hand mike and possibly a hangup bracket, a warranty card, and an owner manual. You may have an extra power cord if your base is for ac (home use) *and* dc (mobile use). Don't throw anything away. Put extra parts like the mike bracket and dc power cord someplace you can remember.

Certain you have everything you should, set the base transceiver in its station location. If you still have to put up an outdoor antenna, don't plug the base in yet. Without an antenna, you won't hear much. And keying the transmitter with no antenna connected could damage it. (You shouldn't transmit anyway unless you already have your class-D license.)

Maybe you bought one of those short rear-apron antennas to use temporarily. With it, you can plug in the transceiver, screw the antenna on, and start listening. Transmitting with the short antenna may not do your transmitter any good. But you can try it. Don't talk for long; it can overheat transistors inside. Keep your breaks short-short until you get the outdoor antenna up.

Erecting an outdoor antenna is more work than setting up the transceiver. Set aside at least three hours for the task. If you get done in less time, congratulate yourself. To aid your antenna construction, the next several pages suggest what you can expect.

You can raise an outdoor antenna alone. But trying alone to secure a heavy, cumbersome, unwieldy, wind-blown structure to a roof might land you in the yard or patio with the antenna on top of you. Get help if at all possible.

In determining where to put your antenna, remember the importance of height. You can legally erect omnidirectional antennas higher than you can beams. Omnidirectionals usually weigh less too, and therefore exert less stress when strapped to a chimney. Your chimney may offer the only logical rooftop support.

Situate your antenna as near your transceiver as is practical. As you use a longer antenna feed-line cable, the effective power radiated from your antenna diminishes. A shorter cable guarantees better performance.

For height, you might stick the antenna up in a tree. But don't lose it all by using an excessive length of coax to reach the transceiver.

Study any supporting structure you plan to attach to. Don't clamp to a chimney with loose mortar. Repair the chimney first. A strong wind takes them both out. Don't fasten your antenna to an unstable post either. You need a supporting structure which can withstand high winds whipping through your antenna.

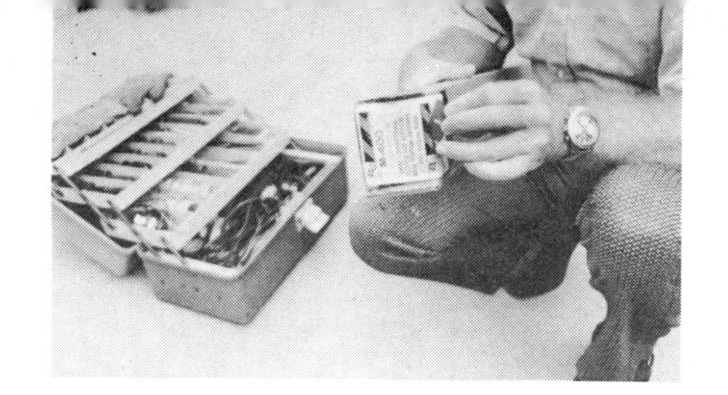

Assemble everything you need before you start construction. Ladder, tool box, coaxial cable, connectors (if your cable doesn't have them attached), pipe for an antenna mast, and of course the antenna parts.

Open the box. Group the antenna parts and check the list to make sure you have everything. Put the screws, washers, and Allen wrench in a can or bowl so nothing gets lost. This is your hardware kit.

Don't lose your warranty card. Put it in your tool box or take it into the house. You don't want it to blow away.

Next, read the instructions through once or twice. Before you begin, be sure you know the nomenclature used for various parts of the antenna; assembly should go easier.

One thing we don't like to say, but must. All too often, instructions for the do-it-yourselfer are miserably unclear; sometimes they're downright wrong and only trial-and-error gets the job done. Some antenna makers do a better job.

We chose for our main base installation the *Starduster* model from Antenna Specialists. And we picked out a 20-foot length of antenna mast pipe to support it. Follow along with us now as we assemble antenna and mast. The sheet told us everything we needed to know. But what we hope to do is breathe a little life into those line drawings that illustrate most antenna instructions. What we show you here, you can apply to those hard-to-comprehend directions that accompany other antennas.

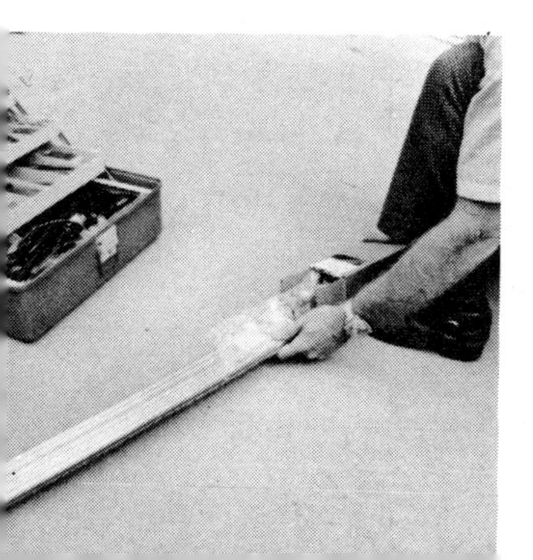

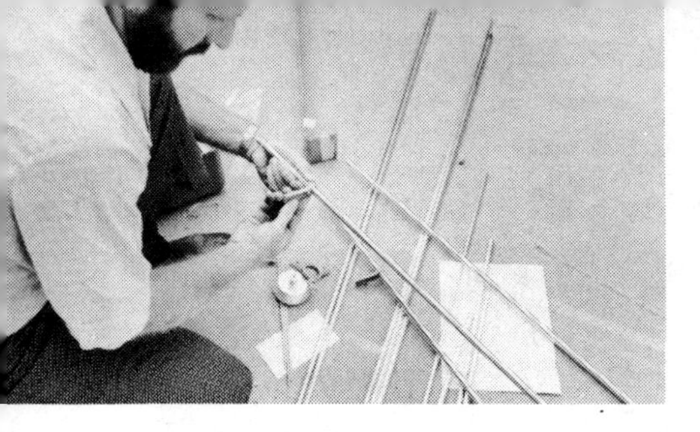

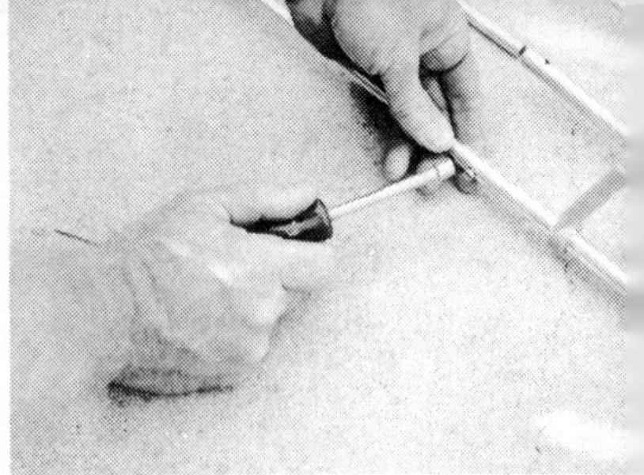

We went by the construction outline supplied with this antenna. So we began by assembling the 107-inch radial elements.

Find the three tubes that have short adapters at the ends. Separate them from the other antenna parts. The tubes are 55 inches long. The short adapters at one end are threaded with a hex nut already on the threads. There's a small hole about an inch from the opposite end.

Extract three small gray plastic insulators from among the hardware. Slip them onto the tubes exactly as shown, so the trough faces the end with the screw hole. If you rush and get careless, you'll slip them on upside down (as our installer at first did here). Later, you would have to undo your work and assemble these radials all over again.

Find three more 55-inch radial tubes. These have a small hole 2 1/2 inches from the end. Also take three self-tapping screws and three large red plastic tips from the hardware kit. Get your 1/4-inch hex-head nutdriver from your tool box. Slip the red tips onto the plain ends of the tubes (the ends without the holes).

Slide together the two tubes, the smaller diameter inside the larger. Match up the small holes near the ends of the tubes. Secure the tubes together by putting the self-tapping screws into the hole in the outside tube. Tighten so the screw penetrates the hole in the inside tube too, making each 55-inch pair into a 107-inch radial rod. You wind up with three of these radials.

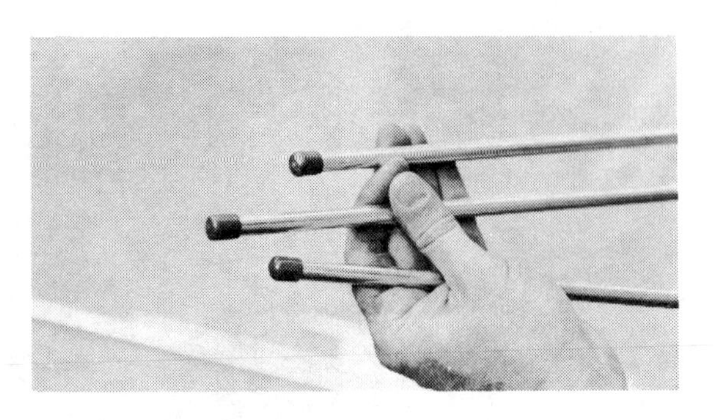

Next you put together the radiator or upper vertical element. It will end up 101 inches long, and takes three tubes to reach that length. You'll find one 51-inch tube with a threaded end and a hex nut. The threaded adapter end is slightly longer than the ones on the three radial tubes. Pick out also the 24-inch and 31-inch tubes. From the hardware kit, take four self-tapping screws, a large red plastic tip, the small hub, and one set screw that fits the hub.

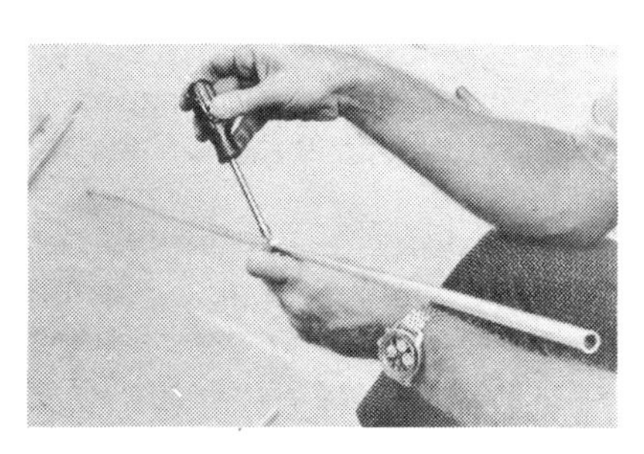

Select the "longest" end of the 24-inch tube—where the drilled hole is 3 inches or so from the end. Slip that end of the 24-inch tube into the drilled end of the 51-inch tube. Align the holes, and lock the tubes together with two of the self-tapping screws.

Now pick up the small hub. One side is flat, the other side tapered. Orient the slanted or tapered side of the hub toward the joint held by two self-tapping screws. Slip the hub onto the 24-inch tube, sliding it toward the larger tube.

The hub will be stopped by the self-tapping screws. Recheck the hub's orientation: tapered side next to the screws. Place the set screw into the hole in the small hub. Tighten it to anchor the hub on the radiator assembly.

Pick up the tube that's a little over 31 inches long. Fit the red plastic tip over the plain end of this rod. The other end has a hole near it. Slide this end of the 31-inch tube into the 24-inch tube. Match the holes, and secure the tubes together with the other two self-tapping screws.

Put the now fully assembled 101-inch radiator aside for a few minutes.

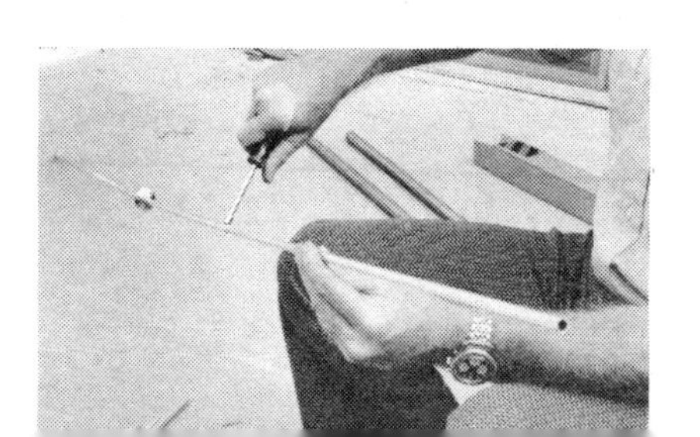

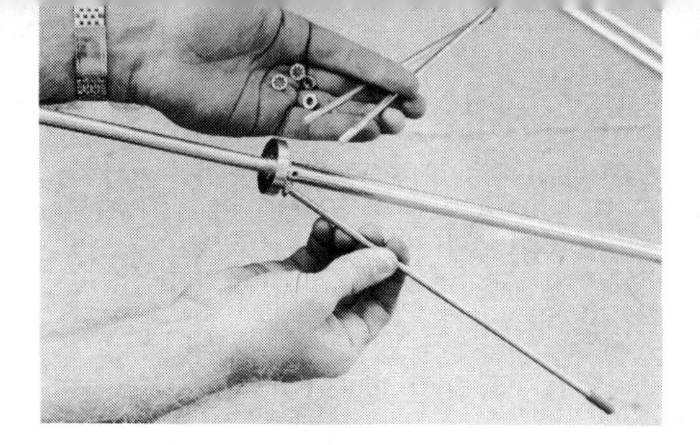

Take the three small-diameter 10-inch radial tubes from the remaining tubes. These radials are threaded at one end. Slip the three remaining red plastic tips from the hardware kit onto the unthreaded ends.

Among the small hardware that's left, you should find three hex nuts and three internal-tooth lockwashers. Holding one of the three 10-inch radials in your hand, screw a hex nut onto the radial first, then slip a lockwasher on.

Prop up the 101-inch radiator assembly so you can work at the hub. You'll see three threaded holes in the tapered side of the hub. Screw the assembled 10-inch radial into one of those holes in the tapered side of the hub. Tighten the hex nuts with pliers or a wrench, to hold the radial solid.

Now repeat with the other two 10-inch radials. First a hex nut, then a lockwasher on the threaded end. Then screw the threaded end into the hub. Finally, tighten the hex nut against the hub to lock the radial there.

Lay the radiator assembly aside again.

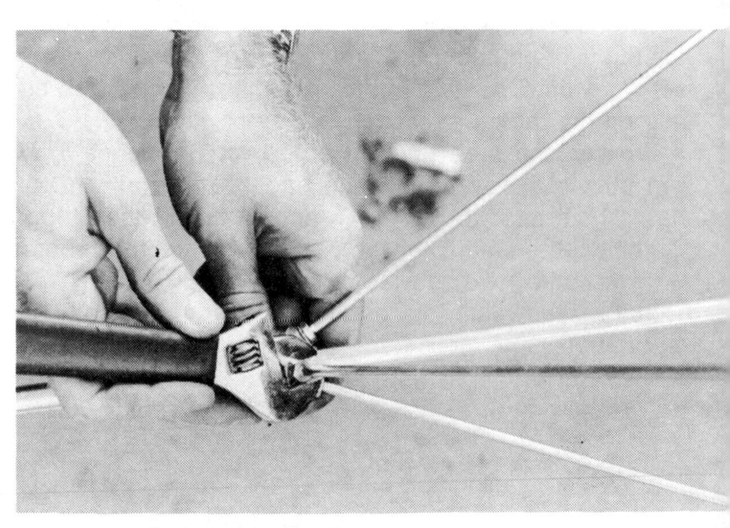

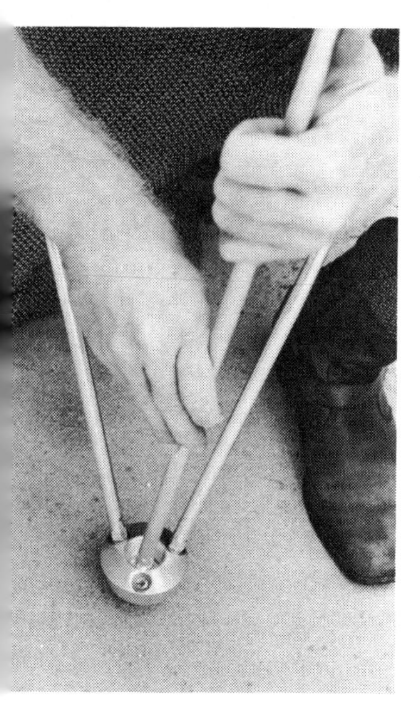

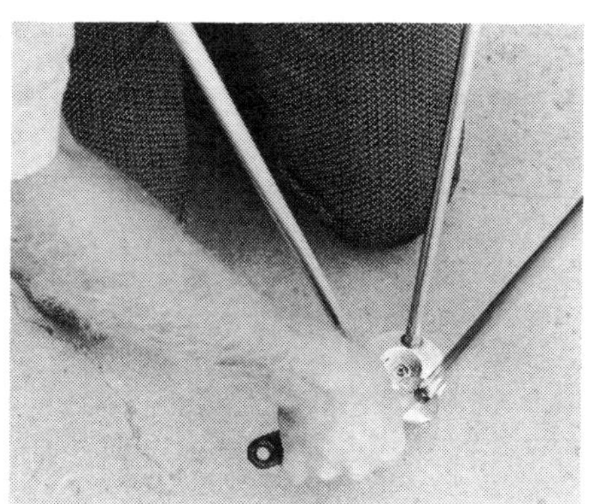

Gather the three 107-inch radial assemblies you put together first. You'll recall they already have large hex nuts on the threaded ends.

Hunt out the large hub among the remaining parts. Pick out from the hardware kit the three large split lockwashers.

Place the hub so the donut side (with the three holes in the rim) faces you. Put one split lockwasher on the threads of each radial, and screw all three radials into the holes in the hub rim. Use a wrench to tighten the hex nuts and secure the radials.

Our installer found that adding the radiator at this point, as the instruction sheet suggests, hampers the work that follows. So from here to completion we digress from the manufacturer's sheet. The antenna gets heavier and less manageable as we go along, so we waited to add the radiator to the hub later. You'll see.

Remember that 20-foot antenna mast pipe? Bring it over. It's time to work with that. Drag out your coaxial cable too, and hope you bought the kind with connectors already in place. You'll also need the three-spoked insulator hub, which has those three radials poking up (or down, depending on how you look at it).

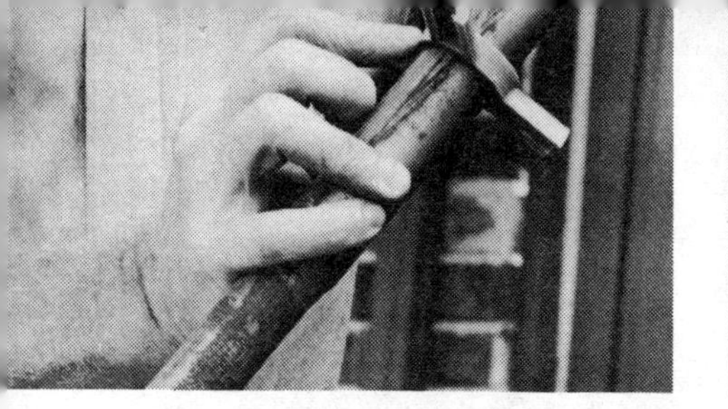

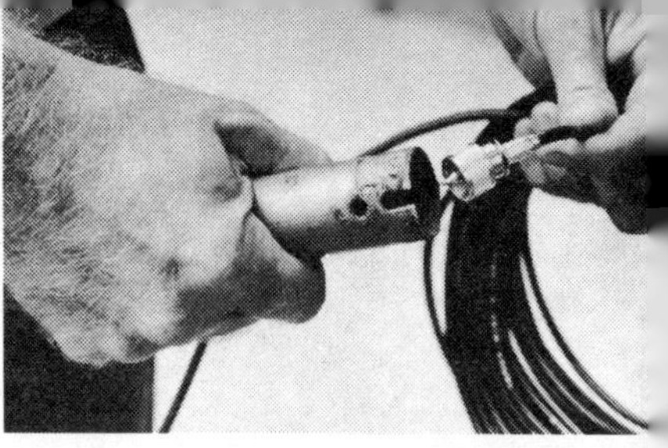

Somewhere among the goodies that are left, you'll spot a spoked insulator. Slide it onto the smaller-flanged end of the antenna mast. Orientation is crucial. The grooved side (under side) should face away from the end of the mast. If the insulator isn't put on the right way, you will get all the way to final assembly and then have to start over from here.

With the cable connector in one hand, and the larger-flanged end of the antenna mast pipe in the other, start the antenna cable through the pipe. The weight of the connector helps pull the cable through the pipe.

If the pipe you're using is in two 10-foot sections, run the cable through the first or bottom section as described. This section should *not* have the spoked insulator on it.

To string the cable through the next section, you must make sure it doesn't fall back through the first. So holding the connector and first section firmly, and with the second section supported on your ladder or lawn furniture, start the cable through the larger-flanged end of the second mast pipe. Let the cable dangle about 24 inches beyond the end of the second pipe before you stop feeding.

Twist the two sections of mast together. Slip the spoked insulator over the cable end and onto the mast pipe, oriented as described above. Tie a loose knot in the cable end so it can't fall back through the pipe.

As you can see, this antenna is now growing by leaps and bounds.

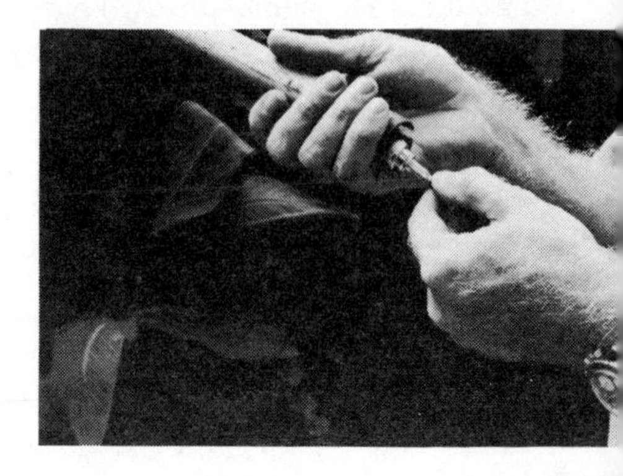

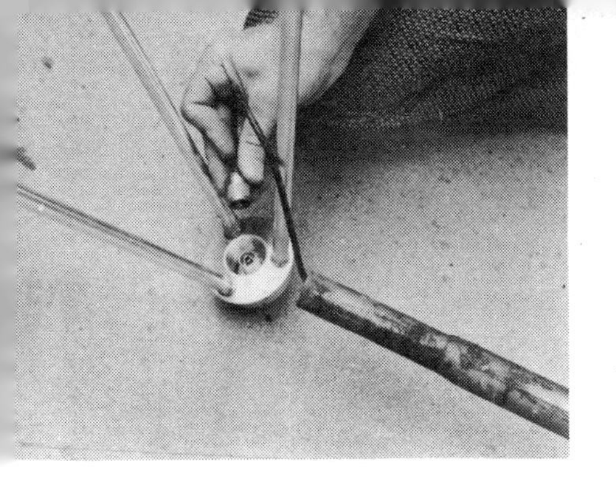

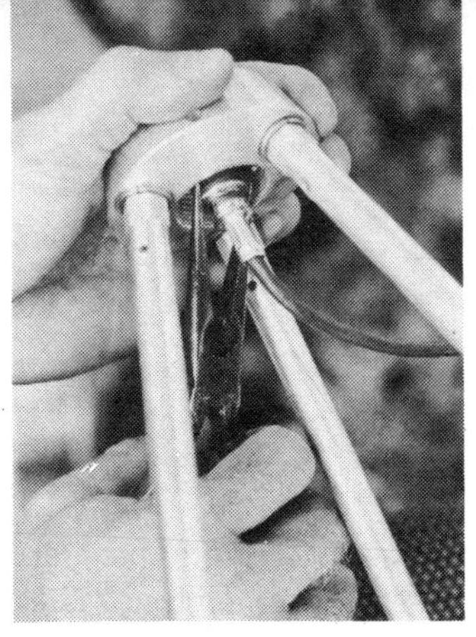

Place the insulator hub on the ground, with the tripod of radials pointing upward. If it's convenient, lean the radials against a shed, fence, or garage or house wall. Bring the end of the antenna mast near the hub, so the cable connector can be screwed onto the connector receptacle in the bottom side of the hub (facing upward right now). Tighten the connector with a pair of longnose pliers. (That's the only tool that reaches the connector to tighten it in these close quarters.)

If you've been wondering why the manufacturer included an Allen wrench with the hardware, fish it out now. You'll need it for the next step.

Fit the antenna mast pipe into the hub. You may have to take up cable slack by tugging it from the far end of the mast. With the Allen wrench, secure the hub to the mast by tightening the three set-screws recessed in holes around the hub. Tighten each one a little at a time, so the pipe ends up centered snugly in the hub. Off-center, it might chafe the cable.

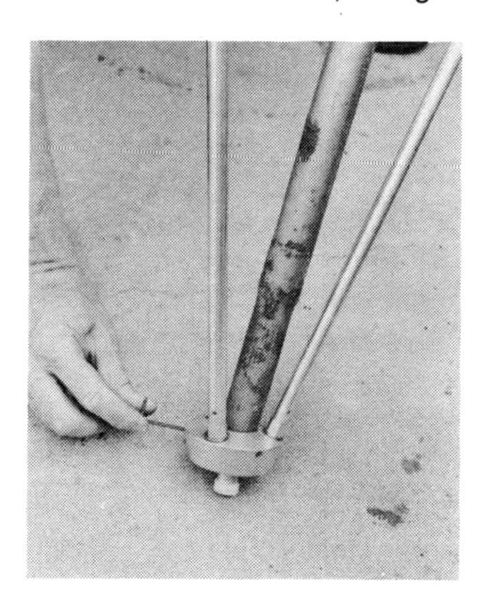

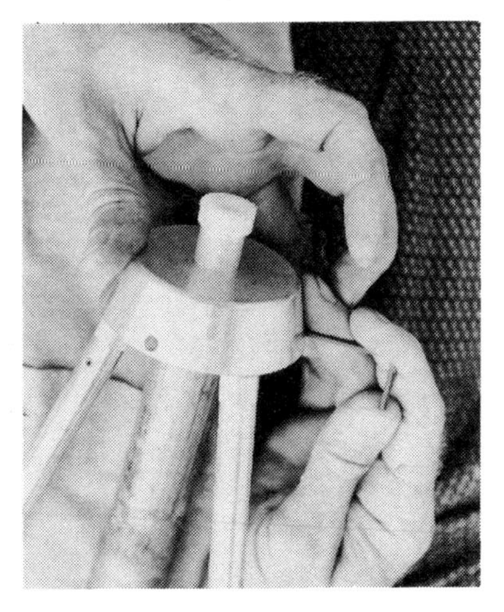

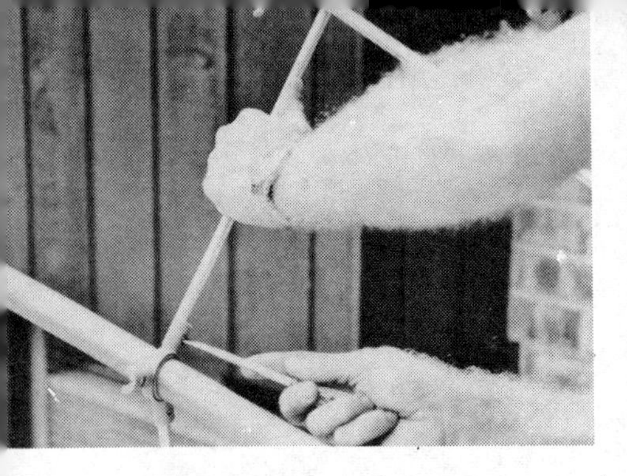

The only antenna parts remaining should be three tubes about 15 inches long. These are braces for the large radials.

You recall the insulators you placed on the 55-inch tube with the adapter threads, right at the beginning? And the spoked insulator you slipped onto the mast just before clamping it into the hub? The braces go between these insulators.

You'll probably have to slide the spoked insulator hub on the mast to match up with the positions of the insulators on the radials. Do that now. Take six self-tapping screws from the hardware, two for each brace. Match up the holes on the braces with the holes in the insulators. Fasten them together by screwing-in the self-tapping screws with your hex-head nutdriver.

These braces stabilize the radials so wind can't wear them out by vibration or break them off. Also, the radiation characteristics of the antenna can't change in windy weather.

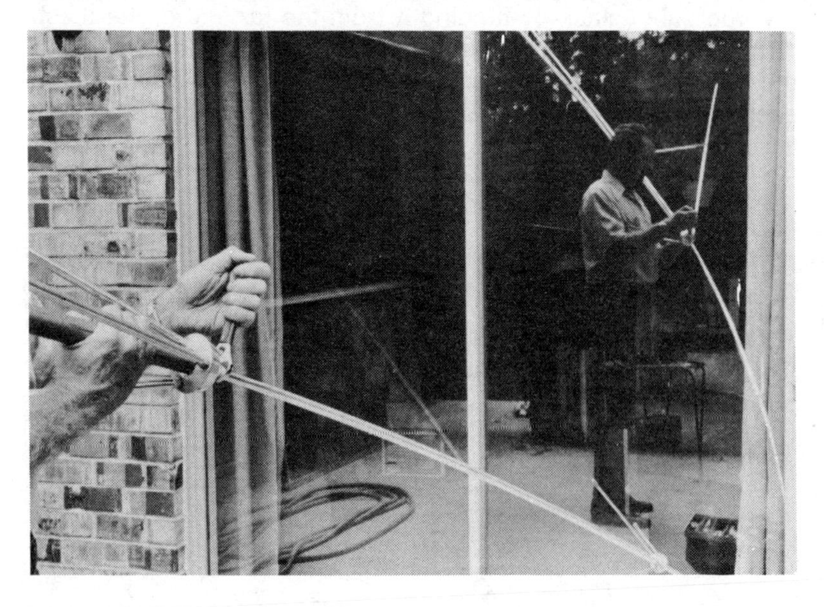

To finish up your antenna installation, you mount the antenna and run the coaxial cable inside to the transceiver.

Once you select where to set the antenna mast, how do you hold it there? You can buy straps that mount the antenna to a sturdy chimney; brackets to brace the antenna against an outside wall of your house; a regular antenna tower; or you can tie your antenna mast against a strong fencepost.

If the bottom of your antenna mast rests on the ground or any solid surface, you'll have to do something to prevent crimping the cable. The better antenna masts include a cable notch along the bottom edge. Your cable must fit in that notch. If it doesn't, the pipe weight can sever the cable or crimp it badly. For pipes without the relief notch, either drill a notch or support the pipe a foot or so above the surface. Hanging the mast allows you to drip-loop the cable between mast and the point of entry into the house.

Never allow sharp angles with your antenna cable. It causes eventual damage in the wire and poor operation from your transceiver.

One more thing. Support your cable every half-dozen feet, by taping it to handy objects. A dangling, flapping cable affects reception, and wears out quickly. It could also be hazardous.

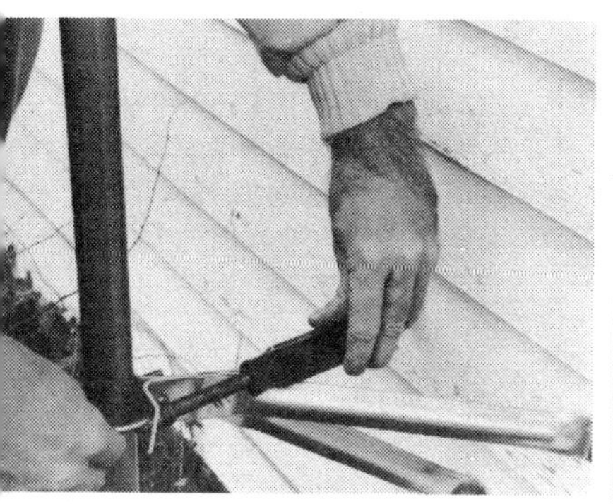

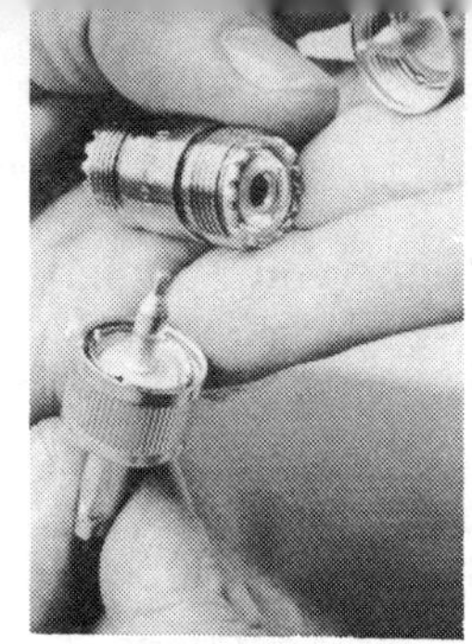

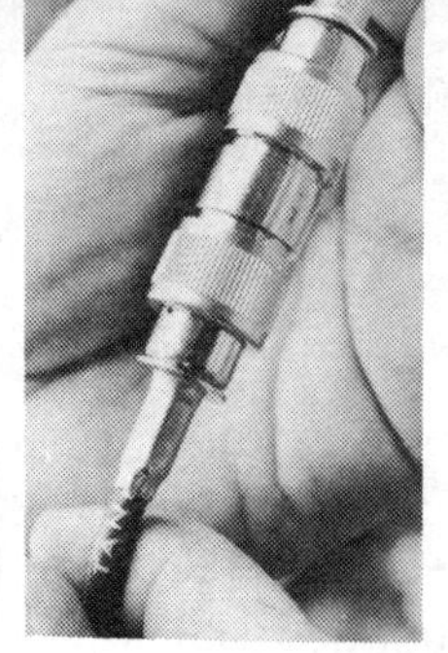

What if the piece of coaxial cable you bought doesn't reach the transceiver? You *cannot* successfully splice coaxial cable. Yet you must somehow lengthen your antenna lead-in.

There is a way. Figure out how much more cable it will take to reach. Add a few feet. Buy that length of cable, or something close, with male PL-259 connectors already on both ends, just like you (should have) bought originally. Or, buy the right length of cable and have connectors put on.

At the same time, buy a PL-258 connector. It is what's called a *female feedthrough.*

Connect the antenna-cable connector to one side of the PL-258. Then connect one end of your new add-on cable to the other side. You've effectively spliced the new cable onto the original.

You finally connect the extension to the transceiver just as you would have the original cable.

Because so many CBers live in apartments, townhouses, or condominiums these days, stringing the antenna cable through a window is common. Also it can be bad, for several reasons. How do you keep out rain and drafts? You cannot close the window or you'll crimp the cable. If you string the cable as shown at the bottom left, rain rolls down the wire and into your home.

Two suggestions: (1) Prevent the rain pouring in by allowing the cable to droop slightly before it passes over the window sill. Rain drips off the bottom of the loop. (2) To eliminate drafts, pick up a roll of weatherproofing tape at a hardware store. Block the window open just enough to accommodate the cable comfortably, and seal the opening with tape. (NOTE: Put a solid screw just above the sash to prevent anyone from raising the window too easily from outside.)

Before you hook up your station, test your antenna installation or have it done. We mention vswr (voltage standing wave ratio) in Chapter 2 and explain it briefly. You or a technician (or a CBer friend who knows how) can measure the vswr of your antenna installation and determine if the antenna radiates well. If it does, then it receives well too.

Most antennas you buy today match your transceiver fine if carefully assembled. But you can misstep in a lot of places, especially in dealing with coaxial cable. You cannot assume everything works just because you were meticulous in construction. Don't take chances. Verify the vswr. Above all, don't operate your station until you do. A high vswr not only produces poor CB operation but if uncorrected can burn up transistors inside your transceiver.

To measure vswr, you need an SWR/wattmeter similar to the one shown here, and a coaxial pigtail. That's a 3-foot piece of RG-58/U cable with a standard PL-259 antenna connector on each end.

In case you have or want to buy a meter, here's how you go about using it to measure standing-wave ratio in your antenna system. Plug your transceiver into the wall socket. Connect the antenna coax to the ANT side of the meter. Connect the pigtail between the transceiver and the TRANS or XMTR side of the meter.

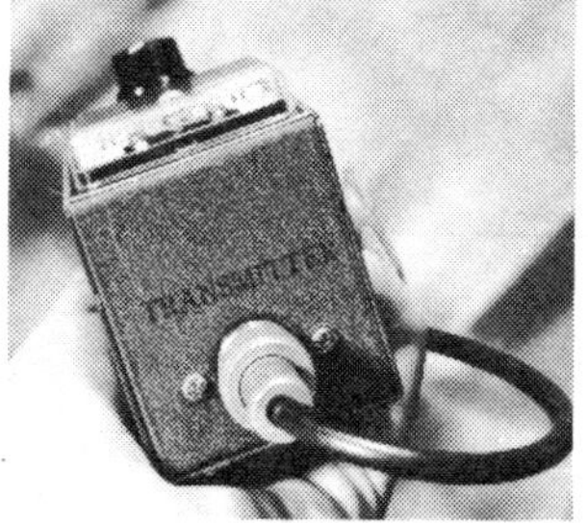

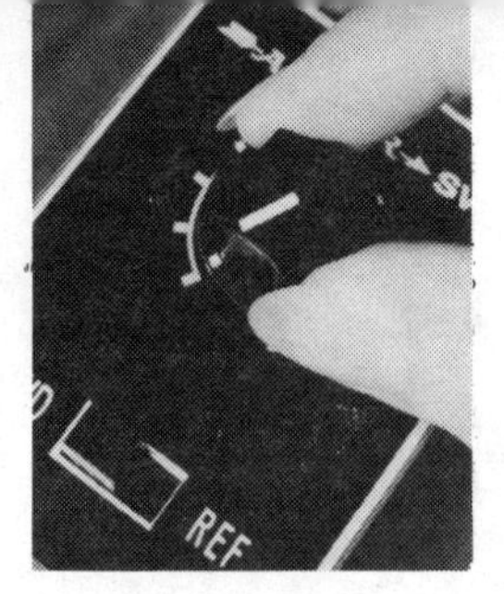

Step 1: Before you fire up the transceiver, turn the meter's Calibration knob to its lowest setting. Then switch the transceiver on, and set the channel knob to 12.

Step 2: On the SWR meter you'll see a switch marked FWD (Forward) and REF (Reflected). Set the switch to FWD. That lets you measure transmitter signal moving *toward* the antenna.

Step 3: Key the mike. In other words, hold down the push-to-talk button. Don't talk. Just hold down the switch, to transmit an unmodulated (no-voice) channel-12 carrier.

Step 4: With mike button pushed, advance the Calibration knob on the vswr meter until the meter needle points to SET or CAL on its scale. Release the mike switch. Do not change the position of the calibration knob now. Leave it turned exactly where it allows a SET or CAL meter reading when you key the mike.

Step 5: Move the FWD/REF switch to REF. Now you have the meter circuits arranged so the meter needle indicates transmitter signal *reflected from* the antenna, back toward the transmitter. This represents signal *not radiated,* and therefore wasted.

Step 6: Key the mike again and read the meter. The meter needle should register at 1.5 or less on its scale.

If the needle indicates vswr higher than 1.5, something's wrong with the antenna system. DO NOT tamper with your transceiver; the trouble is not there. All you would succeed in doing is messing it up.

Instead, go over the antenna and cable thoroughly. Watch for such things as crimped or pinched cable, antenna connectors not screwed on firmly, the antenna touching a gutter or wire somewhere. Worst offenders are connectors.

Check all these points you can. If you still get a high vswr reading (anything over 1.5), call in a CB technician to set things right.

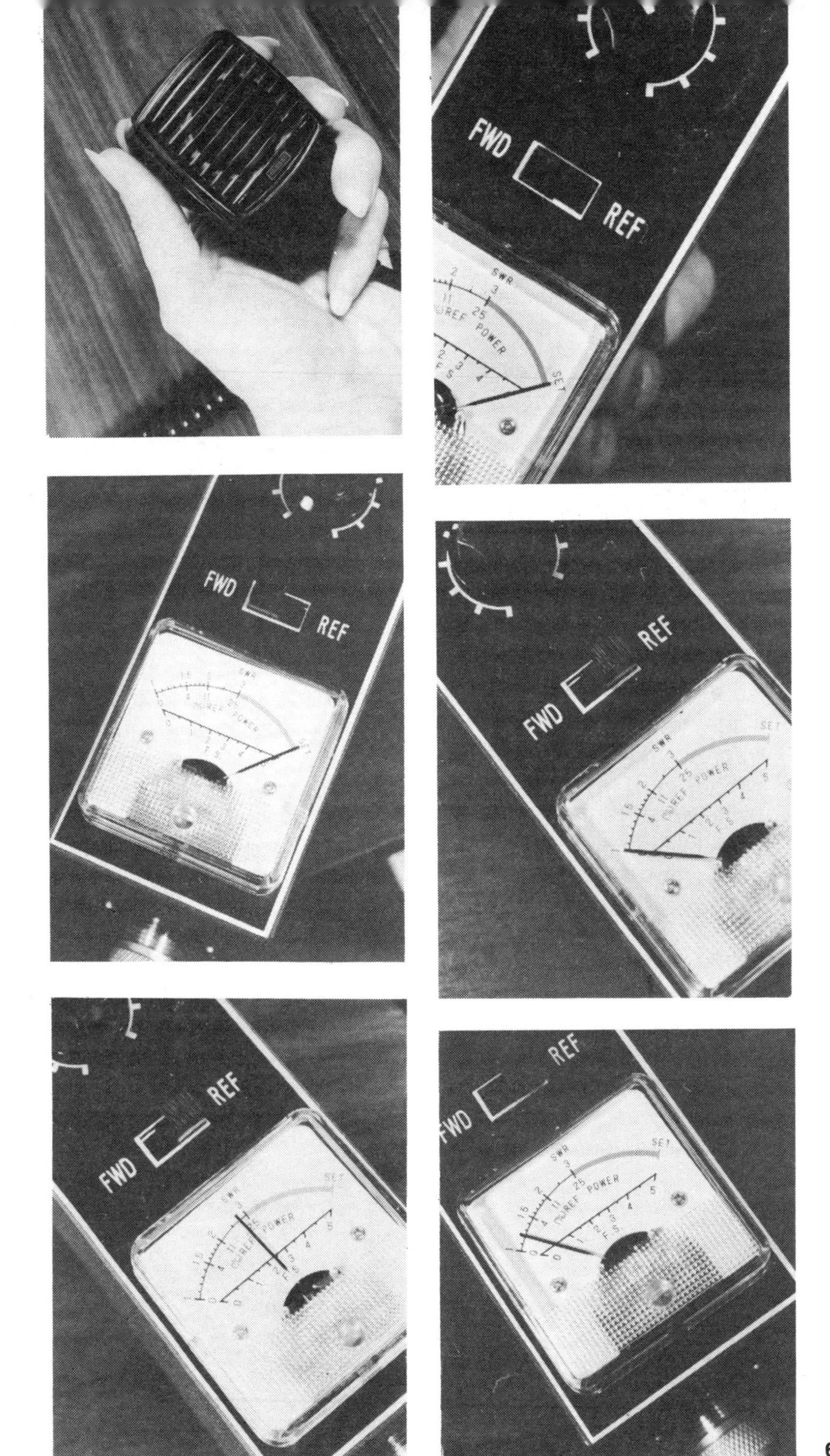
FWD
REF
SWR
REF POWER
F.S.
SET

We don't really recommend that you try putting connectors on antenna cable yourself. Yet, if you already know how to use a soldering gun or iron, and can follow directions meticulously, and will work carefully, you can manage it. Here's how to install a solder-on type of PL-259 antenna connector (the solderless kind doesn't stay together long under repeated use).

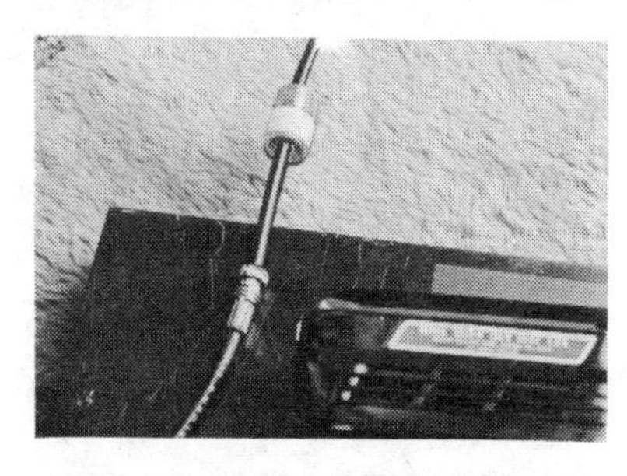

A. Buy a PL-259 connector. If you're using RG-58/U cable, also buy a UG-175/U adapter; it's necessary. Unscrew the coupling ring (backwards) from the connector and slip it onto the freshly cut end of the cable. Slide the adapter on, too.

B. With a knife, cut through and remove the black vinyl sheath, the braided shield, and the inner insulation, about 2 inches from the end. AVOID nicking the center wire, and that's not easy. If you do nick it, cut the cable clean and start over.

C. Next, encircle the black vinyl sheath ONLY, a half-inch back from the end. Twist the sheath to break it loose and slip it off. This leaves a half-inch of braided shield exposed.

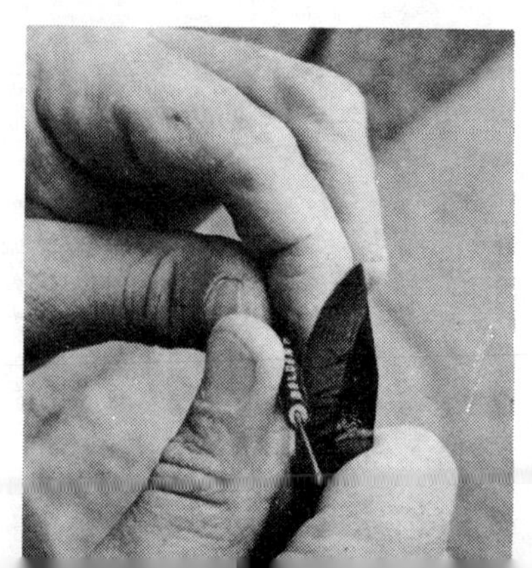

D. Unbraid the shield, and fold the strands back, right where they come out from under the edge of the black vinyl outer sheath. Fan the strands out.

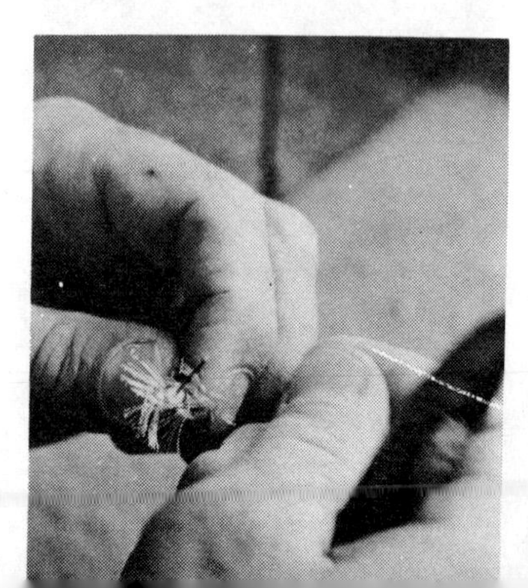

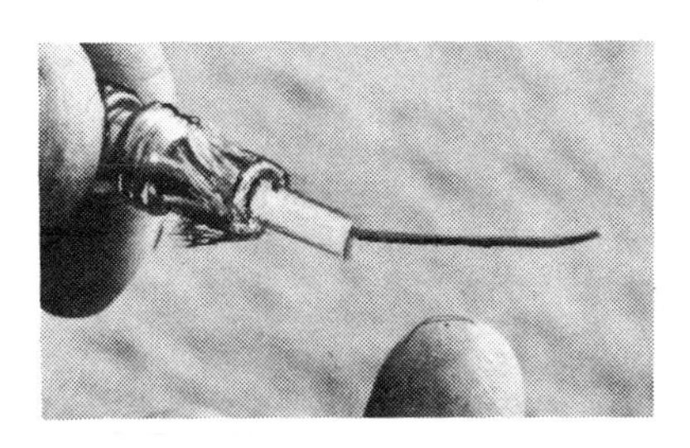

E. Slide the UG-175/U adapter down to the end of the black vinyl sheath, where the strands of the shield begin. Lay all the shield strands back over the tapered portion of the adapter.

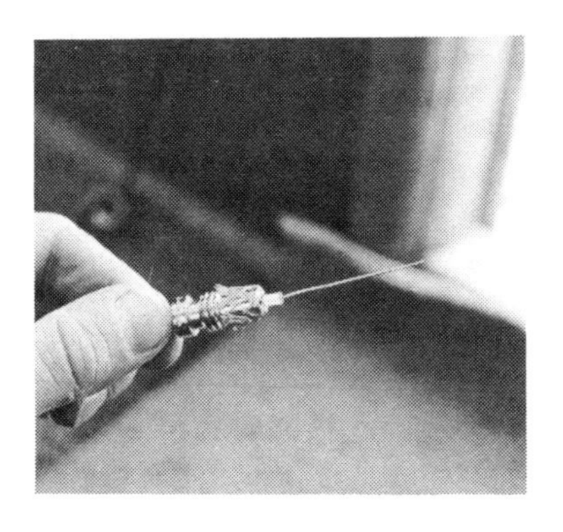

F. Trim off any excess length of the shield strands, so they don't extend into the adapter threads. When you finish this step, ALL the shield strands should lie snug along the taper of the adapter. Clip strands that won't lie smooth.

G. Poke the 2-inch bare center wire into the pin of the PL-259 connector body. Screw the adapter into the connector, carefully so the cable doesn't move inside the adapter. Tighten with pliers. Solder the wire in the center pin.

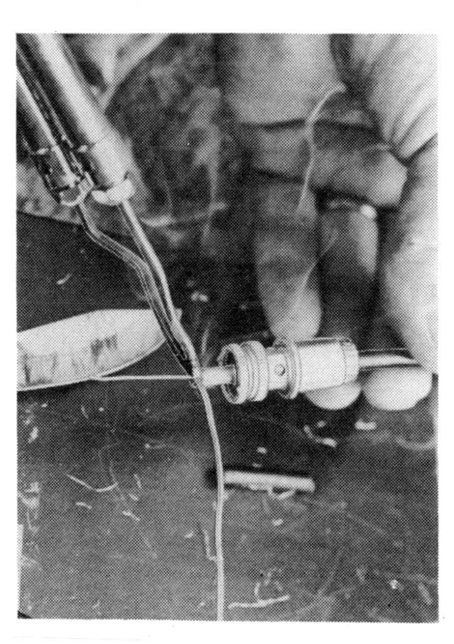

H. You don't need this final step, soldering the shield, unless you have a very hot soldering gun. The adapter makes solid contact with the shield. To complete the connector, screw the coupling ring down onto the body.

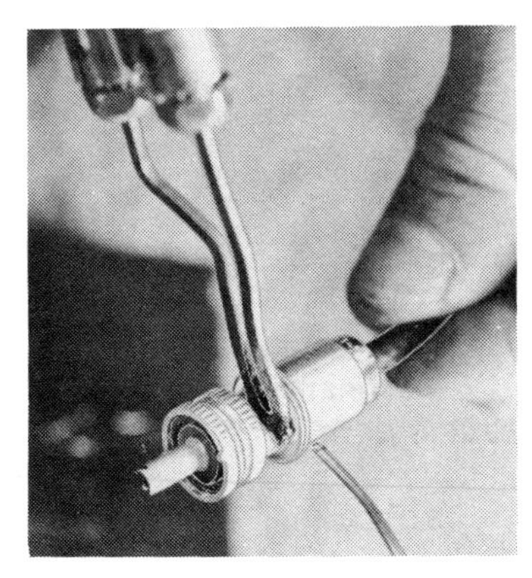

Now that you know more about CB, you'll notice homes with two CB antennas. One is an omnidirectional or monitoring antenna. It picks up signals equally from all directions. However, signals picked up from fringe areas are weak.

One answer to that is a second antenna—a beam. Once contact is made with a distant CBer and the "20" found out, the beam antenna is rotated to point in that direction. The beam, with its concentrated strength, pulls those weaker signals in.

Two antennas feeding one transceiver calls for a coaxial switch. One position couples your monitoring antenna to the transceiver. When someone calls from a distance, you switch to the beam antenna, swing the beam to that direction, and settle down for a crosstown ratchetjaw session.

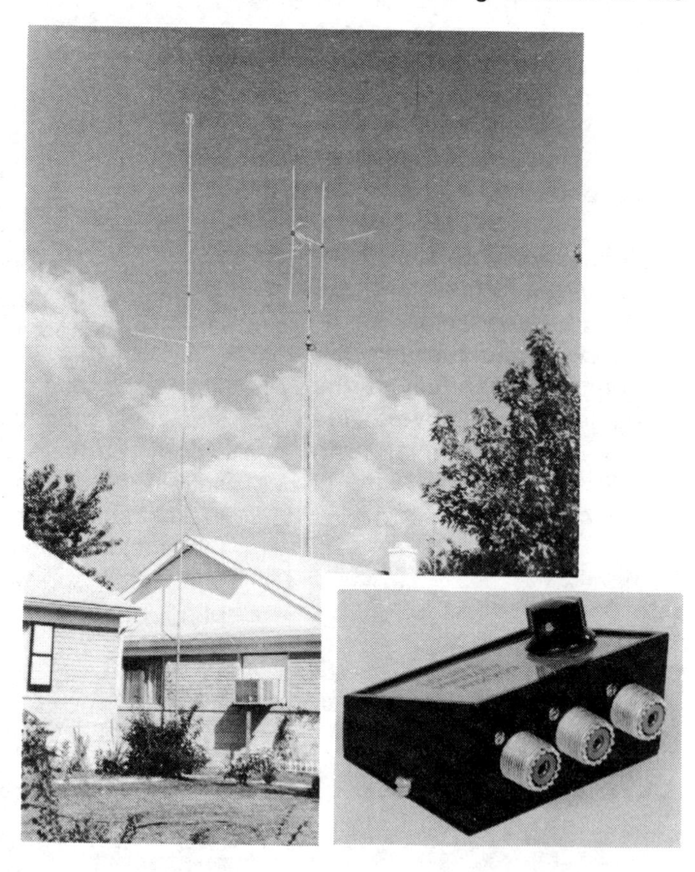

Chapter 5

Using Your CB Radio

Making your first CB transmission brings quite a thrill. Just sitting there at your base station or in your car knowing you're about to join the action can make you as happy as a kid with a new toy. And you can have just as much fun.

If you've had your transceiver set up for a while and have been monitoring the channels, you have a pretty fair idea how to break in. But one word of advice. Don't be disappointed if at first you don't succeed—particularly if you're wanting to contact "Muddy Waters" (who doesn't know you from Adam), the guy you've heard gabbing on channel 15 with other locals. CBers who regularly work a "neighborhood" channel sometimes refuse a "comeback" to a newcomer.

So . . . while you're waiting for your license to come through, invest a little leg work in getting acquainted with other CBers in your area. When you spot a CB mobile parked down the street, and a CB base antenna in the backyard, drop by some weekend and introduce yourself. Explain that you're new to CB. Ask about local CB clubs, if there are any. Visit local CB shops, and let them know you want in with the local group.

Then, when you make that first experimental transmission, you'll find plenty of people out there waiting to jaw a little with you. CBers are generally friendly. You'll find your neighborhood CBer will help you plenty in getting started.

First rule of operating CB: Always open your initial transmission with your call sign. After you "break" for the channel (if it's in use), and get a go-ahead, your initial call may go something like this:

"This is KEQ3427 base looking for mobile unit 1." Wait a few seconds for a mobile CBer to pick up the mike and shout back. If no one answers, try again. "This is KEQ3427 base calling mobile unit 1. Do I have a copy (can you hear me)?" Or you might try, "KEQ3427, Mother Hubbard looking for Easi-Reader. Got your ears on, Easi-Reader?" A handle catches your listener's ears much more readily than your numbers. But you must identify your station. Once contact is made, you can drop the call sign.

If the CBer you want doesn't come back by your second call, key the mike and say, "Negative (no) contact. Thanks for the break. This is KEQ3427, Mother Hubbard at base on the side (standing by)."

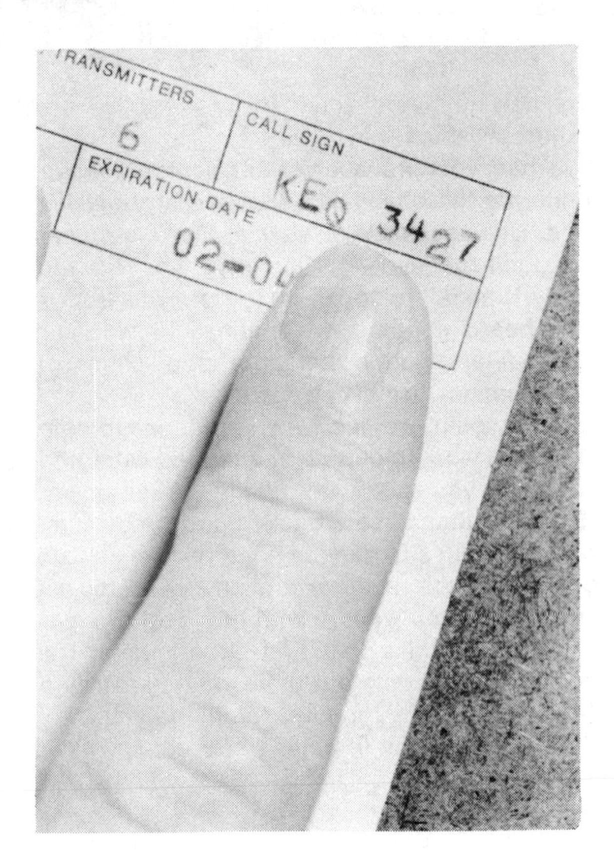

The Federal Communications Commission now allows hobby use of 21 of the 23 CB channels. But that doesn't invite sitting on the channels like you might a telephone. Rules still limit interchanges to five minutes maximum. If you're inclined to be gabby, set a kitchen timer for five minutes. When the bell goes off, back out and clear the channel for other CBers.

In fact, according to the Rules, you must stay off the air at least one minute. If in that time, no one else comes on the channel and you want to resume your conversation, you can. Again, for another five minutes. Then, at the end of that five, break off for another minute. You can continue this as long as your patience and conversation topics hold out. However, it is only common courtesy to allow breakers a chance, if a "break" call comes through (page 26).

Even though the CB channels have been thrown open for hobby and pleasure use, a lot of CBers still use the channels primarily for serious communications. Employees enroute somewhere need to reach the office or shop. Someone at home wants to call a mobile unit for some important errand. Travelers need route information. The five-minute practice benefits everyone. You don't have to talk the whole five minutes. But limit your conversations to five minutes at a time—if for no other reason, just because you're a considerate person.

It's easy to let yourself go, verbally, when chatting away over the CB frequencies, especially after you have struck up a number of channel acquaintances. But one thing you cannot do—even if it's your normal mode of expression—you cannot and must not use profanity. Swearing over the air violates the Communications Act. The FCC will levy a heavy penalty against you if you're overheard. If you do it much, you'll offend enough people to get reported. When the FCC swoops down on you, they revoke your license, slap a smashing fine against you, and could toss you in the slammer for a bit.

So watch your language. There's an even better reason than the legality. Your CB voice goes into a lot of other peoples' homes as well as that of the CBer you're yakking with. Many of those homes have children. CB fascinates kids and they enjoy listening to it.

Nonsensical chatter wastes time and violates FCC Rules. Reports like "Hi gang, I just wanted everyone to know I'm still alive and kicking." The new hobby approval for CB does relax some previous restrictions. But remember, as the number of CBers grows, there's less time per person on the channels. If you're considerate of others, they'll be equally thoughtful of you and your communication needs.

Some other illegal transmissions are whistling, singing or playing instruments, rebroadcasting radio programs or recordings, and keying the mike without talking. All of these irritate other CBers trying to use the channels legally. Anyone who enjoys sending out these unwelcome sounds should be playing in a sandbox, because they're about that childish.

Occasionally you'll hear some joker whistling (so he says) to "warm up his transceiver." Baloney! The way sensible CBers get back at this jerk is to boycott him. When he wants to talk, no one answers him. Finally he gets the message. If he doesn't, he'll be caught eventually by the FCC, and then no one will have to put up with him. Whistling carries heavily on the frequencies, often "spilling over" into other channels. And if this dumbo also has a linear (also illegal) he may blot conversation on channels all around him.

Unfortunately, this type of character probably never read Part 95 (likely he can't read). He doesn't know the Rules and hasn't enough courtesy to use common sense. Seldom does he care whether anyone else gets a chance on the channel or not. He's really just as bad as any CB-transceiver thief, because he is stealing your opportunity to use and enjoy CB radio. Whistlers and carrier freaks (who key the mike but don't talk) can be tracked down easily enough, once the FCC knows about them. And when they are, penalties are stiff.

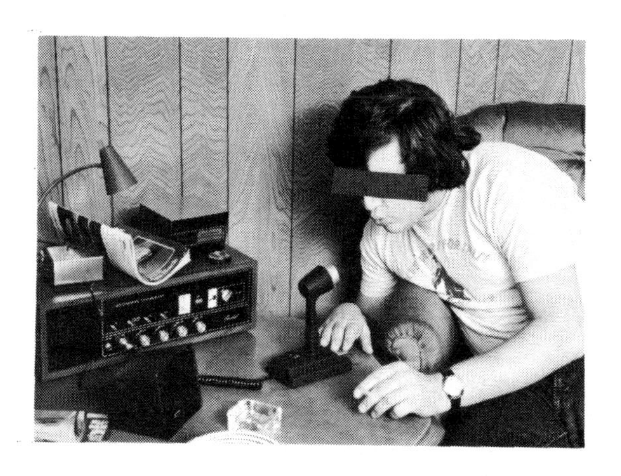

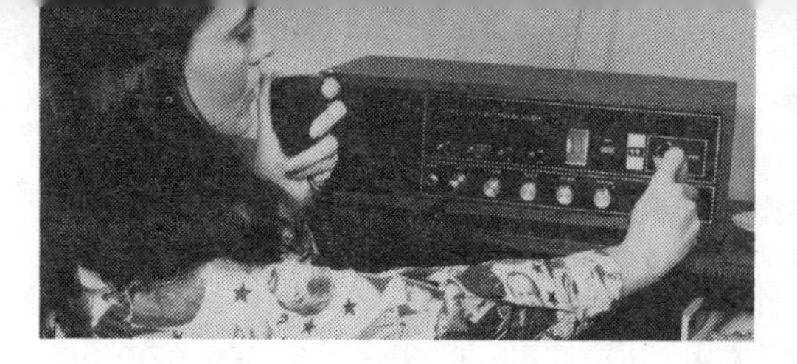

In August 1975, the FCC proclaimed channel 11 a calling channel. The Commission, recognizing the difficulty of making contact over busy general-use channels, set aside one channel (11) expressly for establishing contact. General conversation on channel 11 is now prohibited, the same as it is on channel 9, the emergency channel. So if channel 11 has been your neighborhood "meeting place" for all the local gossip, move to another channel for that.

Here's the way a calling channel is supposed to work. Everyone leaves their transceivers tuned to that channel when they're not talking with someone. Then, say you want to reach Golden Slipper. You just pick up your mike, key it and say, on channel 11, "KZZ9090, Dreamy Sleeper calling Golden Slipper." If she's at her base, she'll pick up her mike and say, "KZX7654, Golden Slipper here. Go ahead, whoever called." (She might not have heard the first of your call, being alerted only when she heard her own handle.)

You come back with, "Dreamy Sleeper looking for you, Golden Slipper. Want to meet on 15?"

"10-4, Dreamy Sleeper. This is Golden Slipper moving to one-five."

"Dreamy Sleeper clearing eleven for one-five."

You both switch. If that channel is occupied, you can either wait a few minutes (less than five, if the other CBers play fair) or go back to 11 and mention another channel. As preparation, you might switch around before you even begin, to see what channel is clear; or you can check some before you switch back to 11 for a second try.

It will take a while for CBers to learn this new system. Once they do, however, they'll find it far handier than the old "break-break" method. And when the FCC adds more CB channels later, the calling-channel idea will prove the best way to handle communications.

Until recently, channel 10 had been unofficially known as a trucker's channel east of the Mississippi River. West of the River, it was channel 19. They might better have been called "highway" channels. Truckers made them popular, and use them to good advantage (pages 16, 22, and 23).

Now, though, truckers have left channel 10 in favor of channel 19 coast to coast. More than likely, you'll hear some drivers on channel 10 for some time to come. And automobiles may move to 11, the new calling channel, and work whatever channels they wish for inter-auto communications.

However this osmosis develops, it must be emphasized that these are not FCC designations, except for the channel 11 calling frequency. Yet, the trucker channel has become a valuable driving-safety tool. Every CBer on the road and many home CBers near major highways participate. Their exchanges have aided traffic flow. For other CBers to clutter these channels with general conversation, when they could just as easily use one of twenty other channels, is a waste. Keeping the trucker channels clear for highway safety marks another way sensible CBers help each other.

Truckers who have (or don't yet have) CB in their tractor cabs will find *Forest H. Belt's Easi-Guide to CB Radio for Truckers* directed to them specifically. It details just about all you need to know about CB technically. And it shows how to make professional CB installations in trucks.

You'll hear some new phrases on channel 19. Things like: chicken coop (weigh station); turkey farm (rest area); bounce (return trip); over your shoulder (behind you); picture taker (radar patrol); funny bunny (disguised radar patrol, in an old car, on a motorcycle, or even on foot, with handheld radar gun); spy-in-the-sky (traffic-spotting plane or helicopter); tattletale (same thing); flip-flop (radar patrol has changed sides of the road); hammer down (rolling at high speed); hammer back (slowed down); and so on.

Just listen sometime. You'll learn CB highway slang quickly. But help by staying off those channels unless you have road information to exchange.

Long before truckers began using channels 10 and 19 (and 17 and 21 out west in some areas) for highway safety, the FCC had officially designated channel 9 for highway emergencies. You may have heard channel 9 called the HELP (Highway Emergency Locating Plan—now defunct) frequency. Channel 9 isn't limited to highway emergencies. You can use it for any local emergency—fire, home or farm accidents, crime reports, tornado or earthquake warnings, a levee break, downed power lines, ruptured water or gas mains, or a pregnant woman going into severe premature labor. Whether you have a home base or a mobile unit, in most localities someone monitoring channel 9 hears you 24 hours a day and will arrange fast help.

Who monitors channel 9? Most often it's the local chapter of REACT (Radio Emergency Associated Citizens Teams). REACT is now a nonprofit organization similar to Red Cross. (It formerly was sponsored in part by General Motors Research Laboratories as a public service, and still receives support from such companies.) REACT's services and achievements have proven successful because hundreds of individual chapters (nationally and in Canada) are held together by a common goal: public service. They are administered by REACT National Headquarters (111 East Wacker Drive, Chicago IL 60601).

REACT members are CBers just like you—CBers who care about other people. REACT members volunteer their time, equipment, and talents for all sorts of local emergency situations. Some teams are well equipped with special trucks, boats, and planes. Others have only their CB radios. Plus a willingness to devote many hours a year to helping others. If you have no special reason yet to thank them, you probably will. Get the jump on that. Join them!

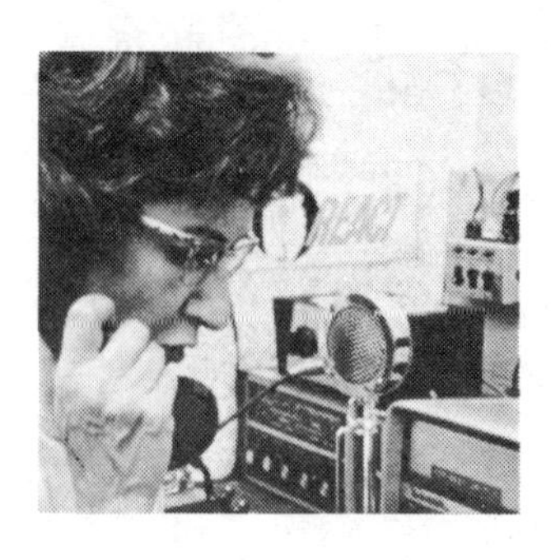

When should you use channel 9? Anytime you're on the highway and you see or become involved in an accident. First, as quickly as you can, warn drivers behind you, using the trucker channel (19). Make sure you give the correct location, to the nearest milepost. Ask for confirmation from someone who will pass the report along. Arrange if you can for another mobile to stop near the scene and keep traffic informed on channel 19.

Then move quickly to channel 9 and report the accident and ask for assistance. A REACT member (or whoever monitors 9 in that locality) starts telephoning for police, wrecker, ambulance, or whatever aid you suggest is necessary. Once these contacts are made and confirmed, the channel 9 monitor may ask you to stay on the channel until help arrives. Usually a highway patrol car can be there within minutes, and the officer will take over.

Stay if the officer wants you to. Otherwise, clear with the monitor, and with your channel 19 helper, before you leave the scene.

You can use channel 9 for home emergencies too. Suppose your toddler gulps down some household cleaner. Radio for help on channel 9 while you try to find antidote instructions on the cleaner label. A REACT monitor who hears you will send help immediately. He'll also phone a doctor or a poison information center and relay instructions of what to do—and what not to do—until help arrives. Some monitoring stations have a phone patch. They can connect you directly with the doctor or poison-information attendant.

You can also use channel 9 for nonemergency highway communications, such as asking directions. Or for something as embarrassing as running out of gas. Many service stations along major highways monitor channels 9 *and* 10, with two transceivers. Even in a city, you can ask street directions. You'll find channel 9 far less congested for this than channel 19.

NOW LET'S LOOK at the mechanics of CB operation. It won't take long to learn to use all of those strange knobs and switches on your transceiver, even if you have one that looks like a space-age computer. Some of these controls, you're already acquainted with, from Chapter 2. But you need to know how to manipulate them.

Start with some of the switches. You may have a switch with CB/PA settings. The *PA* refers to public address. At home, you may never use this function. It requires an external loudspeaker and, on mobiles, often a permit to use it. So, for most operators, this switch never leaves the CB (Citizens Band) position. Yet, if you can't make your transceiver work, check to see if you maybe bumped the switch accidentally.

Your base may have a meter switch. You use the *SWR* setting only when you check the vswr of your antenna system. At all other times, keep the switch in the *RF Power* or *S-Meter* position. Turn back to page 29 to see how to understand the RF Power and/or S meter.

The *CAL/REV* switch functions when you check vswr. You make the test as detailed on pages 68-69. One main difference is, the hookup is already made when you attach the antenna cable to the transceiver. CAL (calibrate) here means the same as FWD, and REV (reverse) is the same as REF (reflected) signal. The SWR SENS knob is the calibration control.

ANL/OFF activates the automatic noise limiter. On some transceivers it may be *NB,* for noise blanker. A few models have both. Their job is to cut down some of the interfering noises you receive from time to time. Whenever you hear no undue noise, keep the switch OFF.

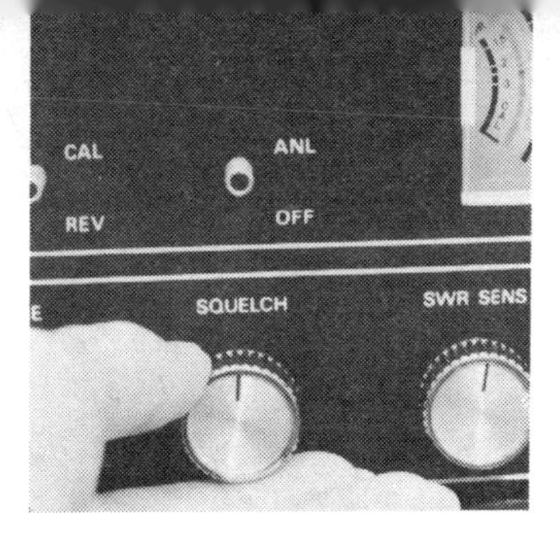

We mention *squelch* and *squelch threshold* briefly in Chapter 2. Here, it's time to show-and-tell how you set squelch threshold. But first, a bit of clarification on what *squelch* is.

When your transceiver is on, and no one is talking on the channel, circuits inside the set create noise that you hear as a hissing or frying sound. This noise is natural. But it could be annoying as you monitor an empty channel. The noise disappears when a signal comes in, "quieted" by the strength of the signal—although you may still hear some of it with weak signals from distant CBers.

A special circuit in the transceiver can block the noise. You can also adjust this squelch circuit to block CB signals that are so weak you can't understand them anyway. Here's how to adjust the Squelch knob.

Find an empty channel, with no one talking. Turn up Volume till you hear the frying. Twist the Squelch control both ways, and leave it all the way at the end that lets you hear the noise. That's minimum squelch.

Advance the Squelch control slowly until the noise abruptly stops. Rotate the knob just a mite farther. You now have set the squelch just inside its threshold. Any CB signal that is slightly louder than the circuit noise will "open" the squelch and let you hear the CBer. Meanwhile, between his transmissions, you won't have to listen to that frying.

Now try another step. Switch to a channel on which you hear extremely weak signals—not anything local or strong. Advance the Squelch knob a bit more, and you can block out those noisy weak signals. Yet, if you don't turn the knob too far, any strong or even medium-strength signal will still open the sound stages so you can hear the voice. (The farther you advance the Squelch, the stronger the signal it takes to open it.) You can thus block out all but the nearest CB stations, if you want to. (Don't leave it too much above threshold, though, or you'll miss calls you may want to hear.)

We've shown you a lot about your transceiver, and we've covered your antenna from tip to connector. We can't leave out your microphone—nor how to use it.

A mike comes with any new transceiver. If the design of your transceiver isn't faulty, the mike that matches your transceiver drives your transmitter to 100-percent modulation. That's all the law allows, and its all anyone can possibly hear at a receiver. More than 100 percent is *overmodulation,* and it sounds raspy and hard to understand at the other (receiving) end.

However, some few manufacturers save costs by providing a cheap mike that delivers less than 100-percent modulation. You get reports of "weak modulation" or "your carrier is strong but I can't hear you too well." That's undermodulation. But before you rush out and buy another mike, try some things to find out if perhaps you're using the mike wrong.

Don't put your mouth right on top of the mike. You're not a rock star or soul singer, and you don't have the same type of mike. Hold a hand mike abou 4 inches from your mouth and to one side. Turn it so you talk *across* rather than *into* it. Keep a desk mike a good 18 to 24 inches away from your face, and on an angle so you don't breathe into it. Use your normal tone of voice.

If this doesn't improve your talk-out quality, then perhaps you do have a faulty or mismatched transceiver and mike.

Shopping for a replacement microphone can prove almost as challenging as picking out your base. You'll find desk-type mikes, amplified mikes, dynamic mikes, noise-suppressing mikes, handsets, omnidirectional mikes, unidirectional mikes, magnetic controlled mikes, ceramic controlled mikes—have you heard enough?

What kind of mike is best? It's up to you and how much money you want to invest. Go on down to some of your local CB shops. Ask questions. Examine store samples. Will you need batteries? (Amplified mikes usually do.) Look at the mike plugs. You won't find any uniformity there, so you'll also have to buy a mike plug that fits your transceiver. Someone will have to install the plug on the mike cord.

If more than one person uses your transceiver, you may prefer a mike with its own volume control. Then you can adjust the mike to produce normal modulation from the sugar-coated tones of "Bunny Body" or from the gruff resonance of "King Kong." You can even raise and lower the head on some desk mikes for tall or small CBers.

Don't be surprised if other CBers tell you that you sound "in a barrel" with an amplified mike. It's not unusual. A partial cure is to talk a little closer, and not quite so loud. (P.S. If you hear someone say your amplified mike "whistles," try new batteries; if that doesn't help, move it farther away from the transceiver. Just a couple of helpful hints.)

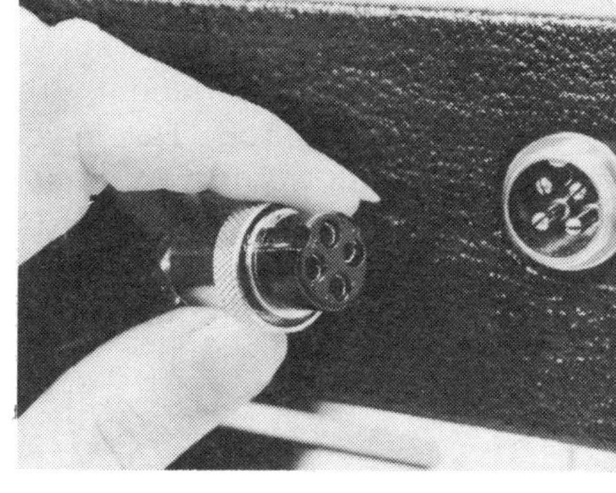

The technique of talking into a handheld mobile mike is as important as how you use desk models—or more important. So note these two big differences.

With any handheld mike, your mouth needs to be close to the mike. It's especially crucial when you're mobile. But, for heaven's sake, don't talk like Broderick Crawford did in the old "Highway Patrol" television series. Jackie Cooper does it wrong in "Mobile One," too. No one could understand a word you said. They set a bad example of mike use for a beginning CBer.

Here's the right way. With microphone almost touching the corner of your mouth, orient the mike so you can talk *across* the ribbed front. Talk in your normal tone of voice; you only distort your words if you shout. Use a level as if you were at dinner, talking across the table. That's all there is to it.

That finishes our quickie course in using CB. Now you have all the basic knowledge to become a CB superstar. So enjoy.

Chapter 6

Putting CB in the Family Car

One primary interest in CB for most people stems from wanting two-way communications in the car. Mobile CBers comprise the majority of all Citizens Band radio users. For every three home base stations bought, CB dealers sell twenty mobile units.

This chapter deals with quick CB transceiver and antenna installation on the family car. We chose this notion of installation, rather than permanent, for three basic reasons. One, not everyone has a lot of time to spend on CB installation. Two, you may not want to get technically involved in complicated antennas and lead-in runs. And three, with CB thievery growing so rife, an easy-in is easy-out, and you can take the transceiver out when you're not in the mobile.

That high rate of CB theft has grown into a serious problem. It's a loss some insurance companies won't cover.

The best way to combat CB thievery? Simply do not leave a transceiver in your empty car. Not even for a few minutes. Many a CBer has learned the hard way. They run into the market for a loaf of bread or a pack of cigarettes, come back out and the transceiver's gone! All a would-be thief has to do is gain entry to your car (fairly easy), disconnect transceiver hot wires (10-15 seconds), and unscrew the coax connector (another 10-15 seconds). You return in 5 minutes and there's no trace of either CB outfit or thief.

Where do you hide your CB then? You can put it in the trunk; if a thief sees nothing visible, he'll go where pickings are easier. You could carry it with you into the store. A lot of truckers do that now when they stop to eat.

It's extra trouble, but take the rig into the house at night. Better a bit of inconvenience than a CB ripoff. Damage to the car may cost more than the CB if a thief is determined and has time, like at night. With your transceiver hooked to a cigarette-lighter plug, removing or hooking it back up is almost as easy as taking the keys out of the ignition switch.

Although quick-mount, quick-disconnect CB installations would make stealing a unit easier, remember that the object is to make it easy for *you* to remove. You'll ordinarily need to connect and disconnect your unit at least twice a day. If you have to fumble around or work at it, you're likely to decide it's too much trouble. You'll just leave the CB transceiver hooked up; and the first thing you know, it's gone.

A CB thief doesn't mind working a little for a free transceiver. If it's a normal installation (bracket-mounted), just let him near it and he's got it anyway. So, the convenience angle is for you, to make the connect/disconnect procedure so easy you won't forget, even when you're rushed.

Here's a fast rundown of what's involved in quick-in/out installations. Details come on the pages that follow.

You connect the two wires of your mobile transceiver to a cigarette-lighter plug; obey the polarity rules on page 98. No need to mount the transceiver in a bracket; put it on the seat beside you. Attach the antenna-cable connector. This sort of hookup takes about 15 seconds to put in or take out.

If you prefer the stability of a bracket installation, buy thumbbolts to hold the transceiver in the mounting. Hardware stores have them. That way you won't need a wrench or screwdriver for in/out later. Time: less than a minute.

You might decide on a slide-mount bracket. The mounting bolts to the floor of your vehicle, and your transceiver latches into the mounting. Some such CB mounts have locks, others don't.

Your local CB store may have additional suggestions. Alarms and plenty of other antitheft devices are available. There are ways to protect against the CB crime wave. If your transceiver gets ripped off, it can be blamed partly on your own neglect, carelessness, or laziness.

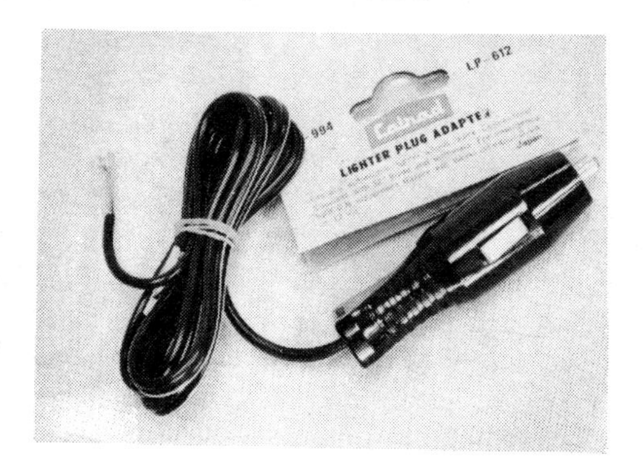

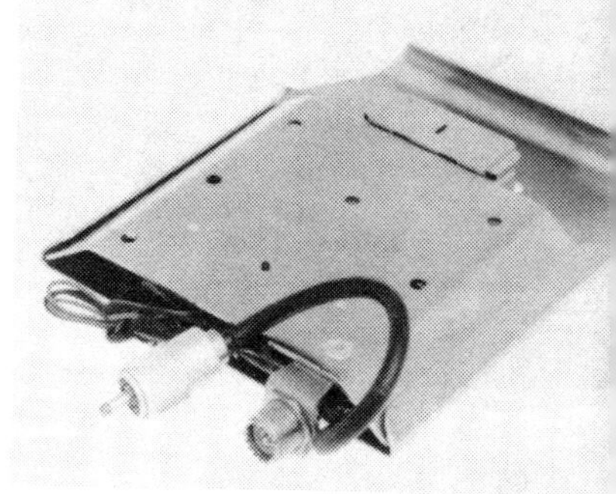

You should perhaps think about mounting possibilities in your car before you buy your CB transceiver. You have several options. With a bench-type front seat, just placing the unit on the seat beside you might do—no bracket needed. But, for reasons you'll see later, this doesn't work so well for transceivers with only one dc connecting wire. You don't need to worry about a mike hanger in the "seat" installation. Leave the mike lying handy on the seat next to you. In some cars with bucket seats, you find space for a CB unit on or beside the console.

The CB-store salesperson probably will carry a model you like out to your car and let you see where it fits. Ask for other placement suggestions.

Underdash (in a bracket or hanger) offers three easy possibilities: below center dash, slightly to the right of the transmission hump, to the left of the steering column. Which you choose depends on how your car is constructed, and on where you seat passengers. A console often blocks a dash-center position. A CB unit beside the steering post must not interfere with operation of your emergency (parking) brake. And no one's knees should bump the sharp corners of a CB transceiver.

Consider another possibility on some cars with bucket seats: the transmission hump in the backseat floor. Worst objection to this would be danger of trying to see to work switches or the channel selector while you drive. The same faults preclude hanging the set on the back of the passenger's bucket seat.

Don't operate a CB in the glove compartment. It gets too hot in there, and some CB manufacturers caution against it. Besides, you'd need an extra speaker, and the mike would be hard to reach and to use.

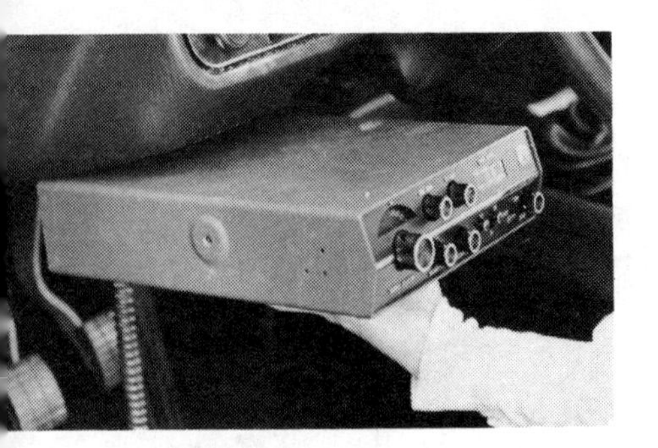

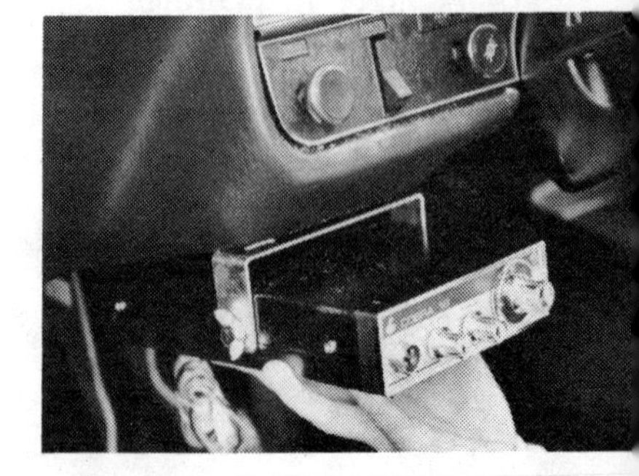

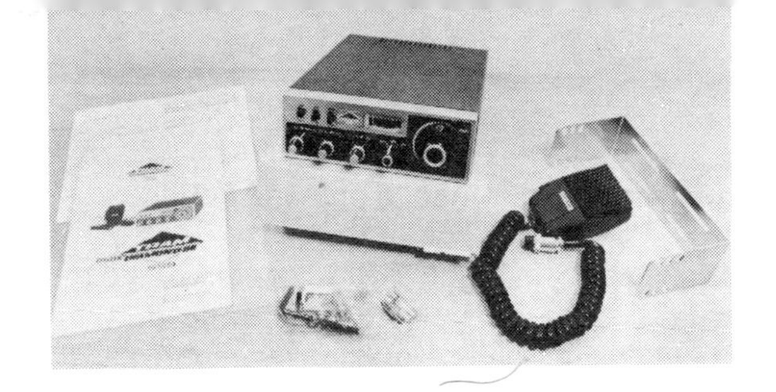

When you buy a new mobile transceiver today, you usually get a mounting bracket, a microphone that matches the transceiver, a microphone hanger, mounting hardware, fuses, mounting instructions, owner manual, and warranty card.

If you decide to use the mounting bracket, you'll need to round up a centerpunch and hammer, and an electric drill and 1/8-inch bit. Depending on the type of screws and bolts packed with the transceiver, you may need a hexhead nutdriver, Phillips screwdriver, or regular screwdriver (or all three). Most hardware packages include an Allen wrench, if the mounting job calls for one. For some antenna jobs, you'll want a tube of windshield sealer, and maybe some miscellaneous nutdriver sizes and a wrench.

One thing before you throw away the empty boxes and cartons. Look inside to make sure you got the owner manual and warranty card. Fill in the warranty card and drop it in the mail.

Keep the owner manual with your station license or Part 95, so you can refer to it easily if you need to. Frequently the owner manual has an electronic schematic diagram of the transceiver's innards. This means little to you unless you're a licensed transmitter technician. You have no legitimate reason to tamper with your transceiver. But should your radio ever need servicing, these schematics help your repairman. So hold on to them.

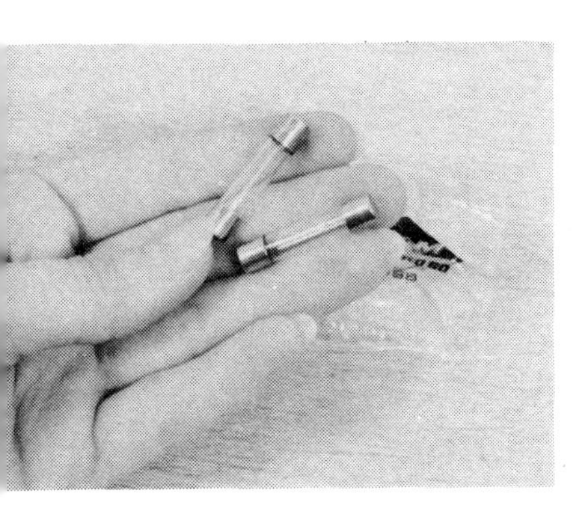

Where you mount your mobile antenna definitely affects communications efficiency.

An ideal location is right in the center of your car's roof But this entails some problems. You have to drill a hole in the roof to mount the antenna. But not everyone has the skill to unhook the headliner and fish the antenna lead down the doorpost properly—and then hook the headliner back so it looks right.

The next best place to install a practical CB antenna is at the center of your trunk lid. You lose a little height, but the surrounding metal gives the whip a fairly good ground plane. You have to drill a hole, but it's in a spot that's not so tough to repair when you sell the car. The antenna cable ends up inside the trunk, an excellent place to start from for a hidden-cable installation.

Fortunately, there's a better trunk-lid idea. It's so much better that it's todays most popular CB antenna mounting for cars. It's called simply a *trunk-lid mount.* You don't have to drill at all. A bracket slips over the edge or rim of the trunk lid, and the antenna mounts right on the bracket. It can go at the center, right behind the rear window, or along any edge.

A less sharp-looking alternative is the trunk-lip mount, in either single or dual. You usually have to drill holes, but in the lip or trough of the car body that surrounds the trunk lid.

Way back, the 102-inch whip mounted on the bumper used to be *the* CB antenna to have. That 102 inches of antenna gives you an entire quarter-wavelength of radiator, an optimum for vertical-whip operation. But bumper mounts have poor grounding, take a lot of time to install, and exhibit an uneven response (radiation) pattern.

Gutter-mounted dual antennas appear increasingly common. You can buy a gutter-mount single too, so don't think you're stuck with duals on the gutter. With any kind of gutter mount, you have a lead-in problem. Most CBers solve it by running the cable in a window, but that's poor. You'll see presently how to run the antenna cable down the gutter and through the car trunk.

Finally, if you want to minimize the number of antennas on your car, and at the same time eliminate that tell-tale "stinger" that thieves watch for, consider a combination AM/FM/CB antenna. It replaces your present car-radio antenna.

PK 145

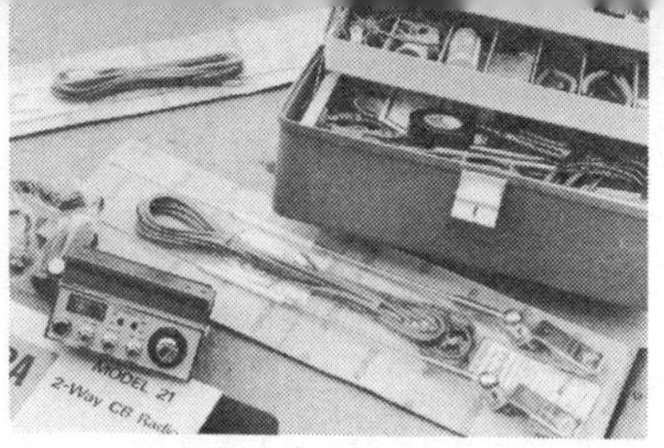

To save cash, you may want to install the CB transceiver and its antenna yourself. You can. It takes a bit of work, if you do it right.

You may have about everything you need in your toolbox. Regular and standard Phillips screwdrivers, a set of nutdrivers, Channellock pliers, adjustable wrench, some small open-end wrenches of the ignition variety, diagonal cutters, long-nose pliers.

Two not-so-common items to buy: a few cable ties and female-type quick-disconnect plugs. You can buy them inexpensively at any large electronic supply store. You'll see later what to do with them.

Open up the CB unit and the antenna you plan to use. Right off, verify that all the little hardware is there. (HINT: Don't open up the boxes over grassy spots. You'll almost for sure lose some of the hardware. Even a magnet dragged through the grass can't ~~always~~ find what drops out of sight.) Should you find any nuts or bolts missing, find replacements at your nearest hardware store before you begin the installation. Nothing seems quite so frustrating as having to stop and go hunt something in the middle of an installation.

Make sure all connectors for the antenna are with it. If the cable end doesn't have a PL-259 rf connector, go buy one. Some antenna packages contain a solderless connector. They work, but are in no way as dependable as the solder-on kind (page 70).

Try several mounting positions. You don't want the unit bumped by your knee or a passenger's. Don't hang the unit on plastic trim around the dashboard. Vibration loosens such a mounting right away. Too, grounding the transceiver's metal cabinet to metal in their car helps suppress some pickup of ignition noise.

Fit the transceiver into a spot where it's easy to reach. You'll have to switch channels, turn the Squelch and Volume knobs, and grab the mike without stretching.

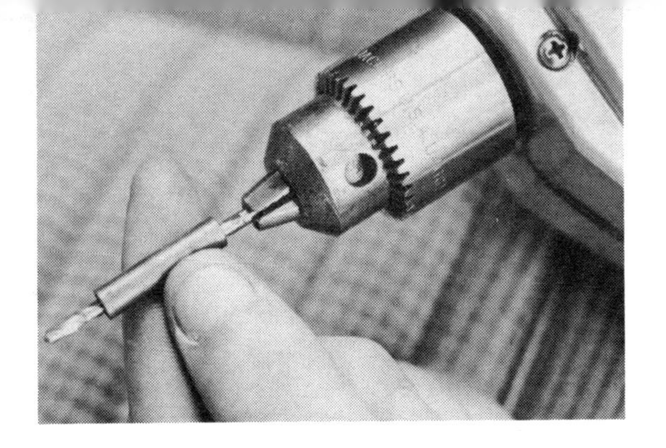

Drilling for mounting screws can mar a neat dashboard. Yet, you can wipe out this worry completely with a few precautions.

Always, whenever you drill on any metal or hard-impact plastic surface, centerpunch first. The slight dimple guides the drill bit and keeps it from skipping around.

A piece of duct tape or masking tape placed over the spot you plan to centerpunch protects even further. Keep the tape in place until you finish drilling. Then peel it off and you'll have a nice clean hole, hardly noticeable when you trade the car off.

Another easy trick prevents drilling into wiring or into braces up under the dashboard. Slip a small bushing onto the drill bit. Length of the bushing should let only enough drill bit protrude to reach through the dashboard metal (or plastic). The bushing keeps the bit from going farther. And of course the tape prevents the bushing itself from marring the dashboard finish.

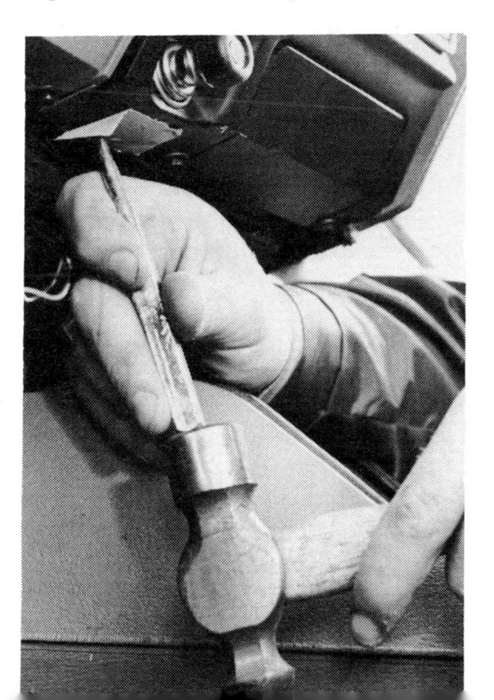

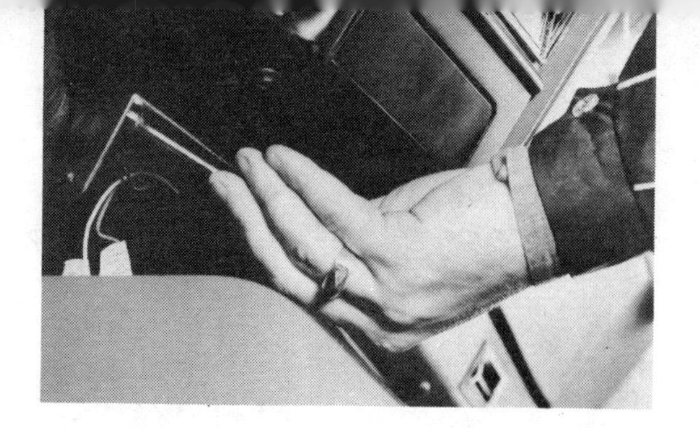

We chose a Cobra model 21 for our sample installation. Its size, in this instance, allowed us to mount it right over the center console in this 1975 Cutlass. Principles are the same for other CB radios and other cars.

With a little observation and ingenuity, you can usually find screws or bolts already in positions that allow you to mount the transceiver bracket without drilling any holes after all. You might have to drill one hole, if the bracket's punched holes don't exactly match a pair of screws already in the lower edge of the car's dashboard. But . . . consider drilling a new hole in the bracket, which would be easier and neater all around than drilling in the dashboard.

Again, however, we should caution against mounting the transceiver bracket to plastic trim. The bracket must screw or bolt to metal. It's important for mechanical solidity and for thorough electronic grounding. If mounting bolts are hard to reach with regular nutdrivers or stubby nutdrivers, try a small open-end wrench. Just be sure the bracket gets solid mounting.

Hang the transceiver in the bracket. Don't fasten it too tight yet; tightening with a quarter instead of a screwdriver will do. You'll have the transceiver out again before you're done. You can tighten it later. Also, you may want to use thumbolts (for easy in-and-out) to make the final fastening.

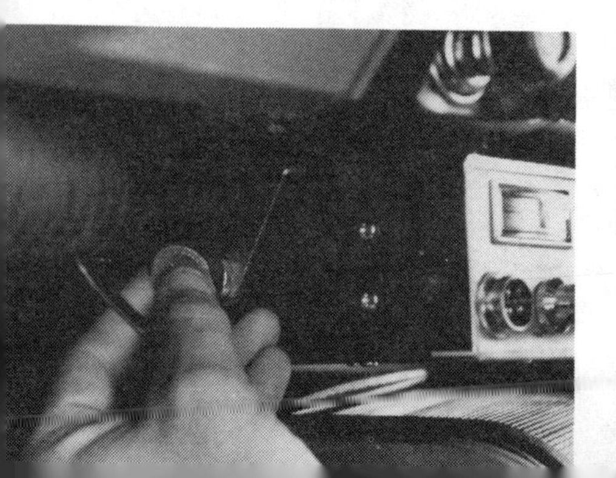

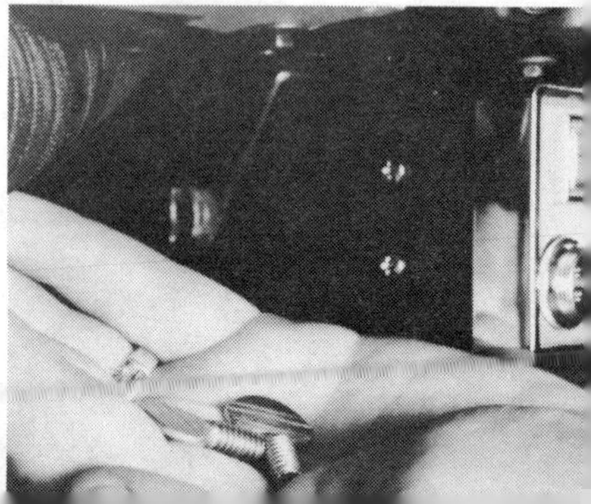

Before you can connect up the transceiver, you have to find a convenient source of 12-volt dc. Normally, the most accessible spot is the fuse block. It's located differently on various makes and models of cars, but your mechanic can point it out to you if you can't find it. Your car owner manual should show where it is.

Typically, you would use a voltmeter to find which vacant terminal posts on the fuse block are "live"—that is, have voltage on them. Make this test with the ignition switch turned to ACC (accessory). If there's an empty fuseholder in the fuse block, slip a 5- to 10-amp fuse into the slot. That should give you voltage at the empty terminal post.

Set the voltmeter for some range above 12 volts. Actually, you may read nearer 13 or 14 volts on the meter, if your car's battery is up. Clip the negative (black) test lead to a ground point, and the positive (red) probe to the terminal post on the fuse block. No reading, no voltage. Normal reading: you're ready to proceed.

All modern cars supply 12-volts dc or thereabouts. One other factor about dc voltage is important, though. That's *polarity.* American-made automobiles, and most from overseas too, have the negative battery post connected to ground—the frame of the car. The system is called *negative-ground.* Only in trucks do you find positive-ground systems.

Polarity is important to CB installation. You must be careful to hook up the CB transceiver in the right polarity. The majority of today's transceivers have two wires for the dc voltage. One is black, the other red. Besides this standardization of colors, you'll find a tag on the hookup wires, reiterating which wire should go to which battery polarity.

Black goes to negative; red goes to positive. This means that, given the typical American car, you connect the red dc wire to the terminal post on the fuse block. (In a positive-ground truck you would connect black to the "hot" side of the dc circuits.)

Typically, the positive or red wire has a fuseholder and fuse in it. The fuse is small, seldom more than a 2-amp. It protects the dc power circuits in case of a short circuit inside the transceiver. The fuse in the fuse block is usually too large.

Once more, before you make any connections: (1) Be certain your car or other vehicle has negative ground. (2) Make sure you have identified the dc wires of the transceiver correctly.

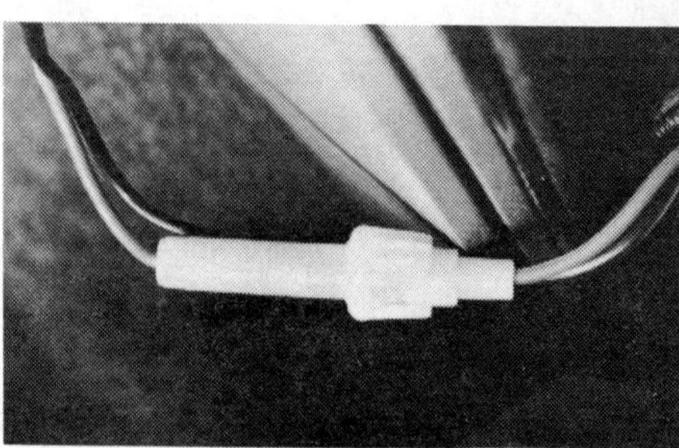

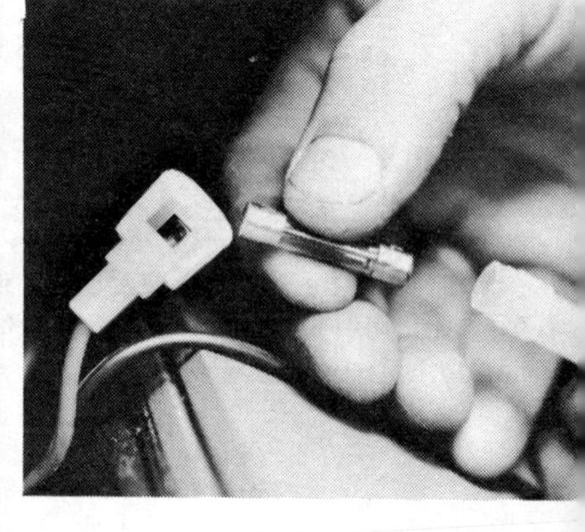

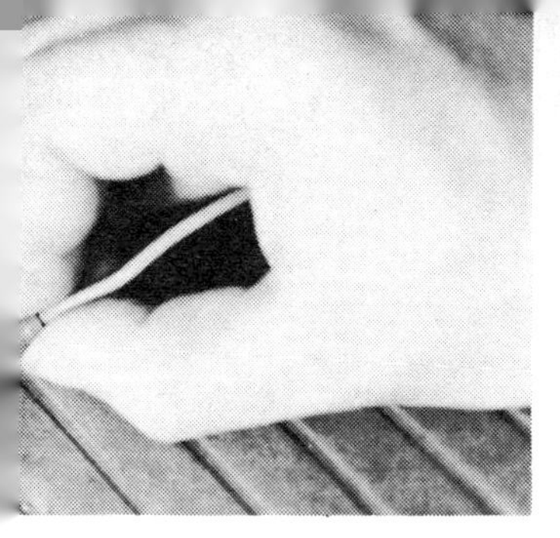

You can buy *quick-disconnect* lugs at most electronic or auto supply houses. You want the female variety, to mate with the male terminal post on the fuse block. The 1/4-inch size fits most cars. Buy the solderless kind, and you can crimp one onto the red dc wire from the transceiver. (Photos on the next page show how to crimp. A crimping tool costs only a couple of dollars.) To make the power connection, then, you merely slide the disconnect onto its mate on the fuse block.

Sometimes you can't find an empty terminal on your car's fuse block. A tap-in, sometimes called a three-way connector, makes a convenient and neat—and also safe—way to splice into an already hot wire. The wire that runs to your cigarette lighter is convenient and easy to identify. But any hot (voltage-carrying) wire will do. Two disadvantages accrue: you can't always tell which wires are hot; and you may hook into a non-switched wire, which means your CB radio stays on even when the ignition switch is off. (This might not be a disadvantage, but you have to decide that for yourself.)

Finally, consider the cigarette lighter itself as a power source. If you want the transceiver easy to take out, there's the best answer. You can buy cigarette-lighter plugs at most CB shops and electronic supply stores.

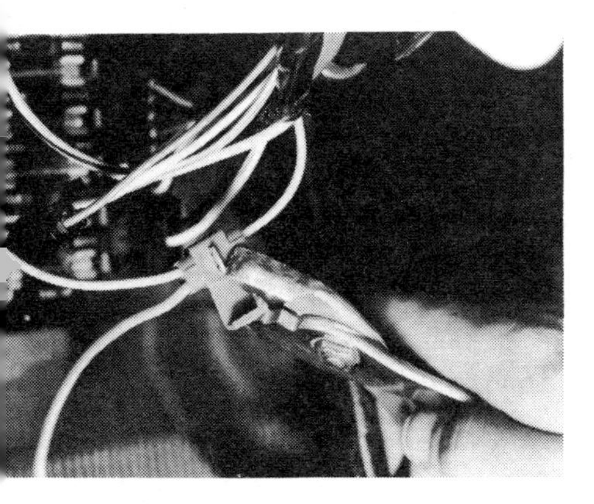

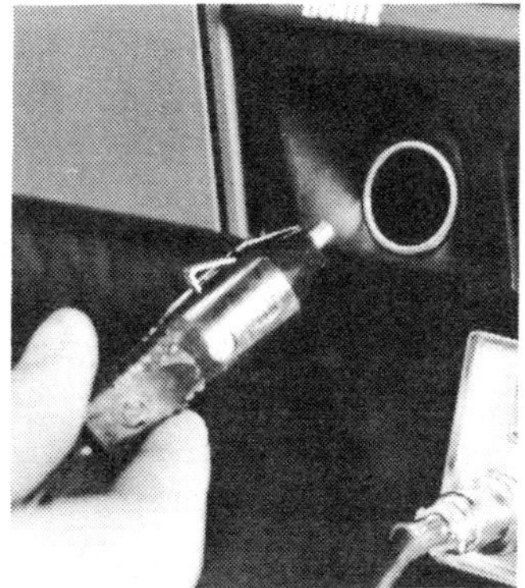

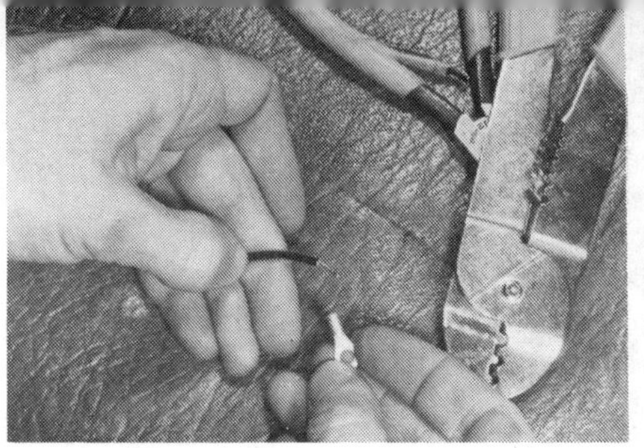

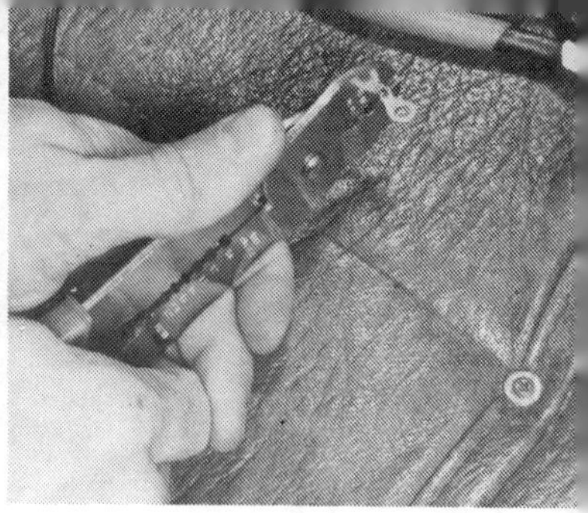

Here's how to crimp a lug onto a wire. Trim the insulation back, exposing about 1/4 inch of bare wire. Twist the strands together. Poke the wire into the barrel of the lug or quick-disconnect terminal. Set the lug barrel into the crimper and squeeze the handles of the tool. The result is a solderless and very solid connection. Lugs are ideal for any wire connection you have to make.

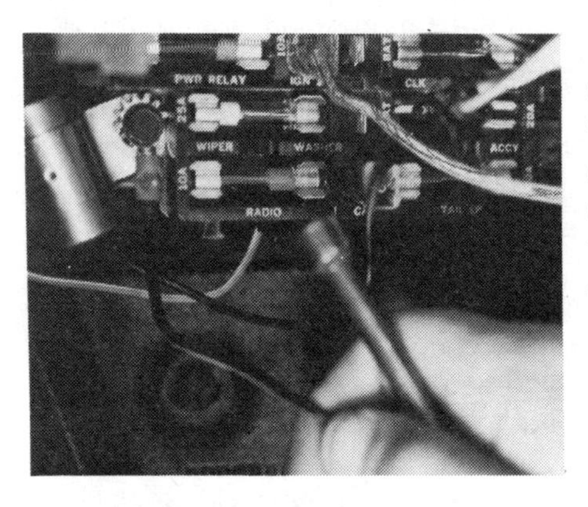

In negative-ground cars, you connect the black transceiver wire to ground. Best way to ground a wire is to crimp a lug onto the end, as shown above, then find a screw that makes solid contact *into metal*. That's important—that the screw go into solid metal that connects ultimately with the frame of the car. Without that connection continuity between the black transceiver wire and the car battery (the negative battery post goes to the frame, you remember), the transceiver operates not at all or very poorly.

One-wire transceivers need a ground between the radio cabinet and the car's ground system. You may have to add a piece of heavy wire or metallic braid. You'll know when the transceiver needs this *bond* connection: You can't turn it on unless the antenna is connected. That situation MUST be eliminated. You cure the problem by crimping a lug onto each end of a solid piece of metallic braid or heavy wire; fastening one end tightly under a screw on the transceiver chassis; and tightening the other end under a screw that screws into metal.

Install the microphone hangup bracket near enough to the transceiver that the cord doesn't stretch. And of course hang the mike so you can reach it easily from driving position. You'll find a dozen good places in almost any car, all of which meet these two criteria.

If you have to drill holes for mounting screws, observe the tape/drill-bit-bushing/centerpunch precautions mentioned on page 95. Be even more careful, because scars made mounting the mike hanger generally show for all the world to see. You don't want a messy car dash, since you're making your CB installation so neat otherwise.

You can even hang the microphone on the side of the transceiver. Just be sure you orient the hangup bracket so the mike can't fall out, and so you don't have to look away from the road to pick up the mike or hang it up.

From the photos on the page facing, you can see the real work of putting on a gutter-mount dual lies in running the antenna cables. Dealing with two lead-ins, one for each antenna, brought together in a common connector at the transceiver, institutes some considerations you might not have thought of.

Anytime you plan to install dual antennas, manage the cable run first. Why? Because you should—indeed, you must—begin at the transceiver and not at the antenna.

Lay the transceiver near, but not in, its mounting bracket. Screw on the connector that terminates the pair of coaxial cables. Separate the two cables. Drape one toward the left side of the car, and one toward the right.

Dress the cable up under the dashboard. Find some wiring already well secured there, and fasten the cable so it can't drape down and tangle a passenger's feet. *Cable ties,* such as the one pictured, work best for this. Wrap the tie around the antenna cable and the wiring harness (or air conditioner hose) you're anchoring to.

Some cable ties have a small hole on the end. This facilitates anchoring the cable in the door facing, so it can't flop around and get pinched in the door. Wrap a cable tie around the cable, paying careful attention to which way the strap turns. There's an anchor screw right in the door facing, holding soundproofing material. Remove that screw, poke it through the cable-tie hole, and reinstall it.

Next, run the cable up the gutter/drip channel alongside the windshield. On some cars, the drip gutter or trim fits tight enough to grip the small cable.

Clamp the antenna to the gutter. Then connect the cable. Slip the center-wire lug on the antenna-rod bolt first, but don't tighten the nut yet. Install the grounding hardware, which clamps around the cable shield where it's bare just back of the end. Finally, tighten both.

Smooth out the cable, adjusting it in the gutter so none sticks out. You may have to work a little of it back down into the car. That's okay. Just coil up any excess cable and stuff it down behind the kick panel just forward of the front door. Close the door a couple of times to make sure there's no pinching. Then "lock" the cable in the trough with a couple of spots of windshield sealer.

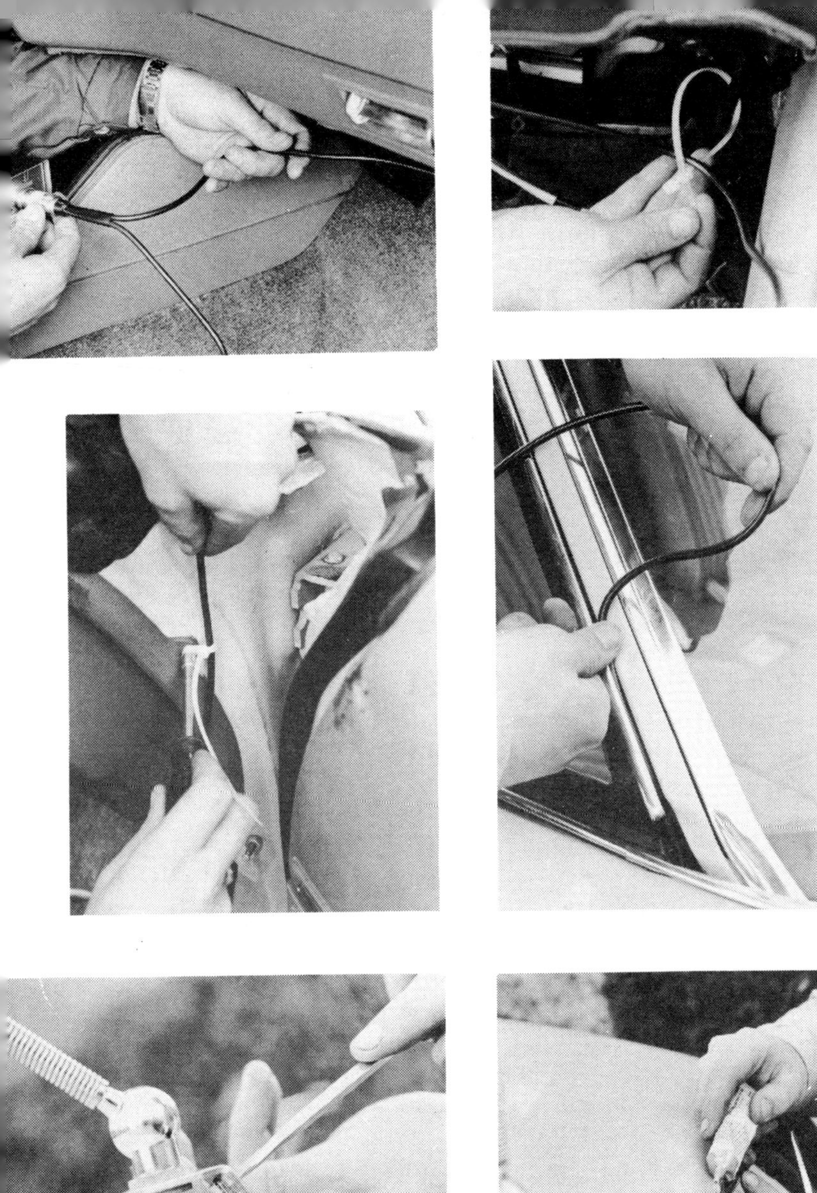

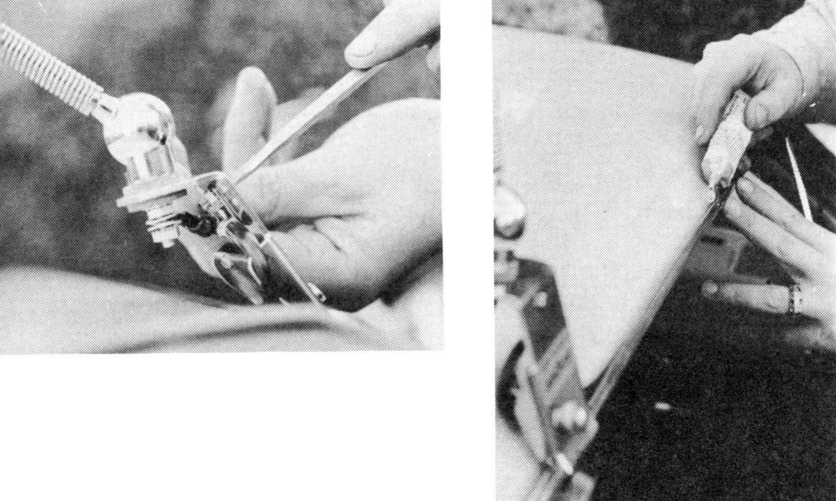

Trunk-lid antennas go on easy too. Yet, even there, we suggest you make the lead-in run first. Where you begin depends on the cable supplied. Some have the connector already on, and you connect the free end to the antenna when you get it there. Other models have the cable fastened to the antenna already, and you add the connector after you feed the cable through the car to the transceiver.

For our sample installation. We chose a New-Tronics model XBLT-4 mobile CB antenna. What attracted us is the cable. The transceiver end already has a PL-259 connector soldered on. At the other end, the cable fastens permanently to the antenna. This simplifies mounting on the trunk-lid edge.

Yet, you have to be able somehow to fish the cable through from trunk to driving compartment. A foot or so from the antenna, the maker has cut the cable. At the break, you'll find a connector of the type used for auto radio antennas. It's called a Motorola pin-type.

With this antenna, we prefer to begin at the base of the rear seat. Take the seat out. Older-model cars have a hook-type retainer; you push the seat backward and lift the front edge. Rear seats in newer cars are held by two bolts, one near each end. Try a 7/16-inch nutdriver on them.

Your task is to poke the pin-plug end of the antenna cable up through the bulkhead between trunk and rear seat. In some cars, it's easy. In others, you may need to "fish" from the trunk. To do that, straighten out a wire shirthanger. From the trunk, poke the hanger wire down until it comes into view inside, near the floor under the rear seat. Bend a small hook and engage the pin plug. Gently, so you don't tear off the plug, draw the antenna cable up and into the trunk. It's easier than it sounds.

Now go back inside the car. Remove the splash trim from the passenger side. A few Phillips screws usually release it. The carpeting edge may be held down with sticky gunk, but lift it up anyway. Run the connector end of the antenna cable underneath the carpeting, all the way to the front. Up front, work the cable around behind the right-side kick panel.

Connect the PL-259 to the transceiver. Use cable ties to support the cable up out of the way. Work any slack over behind the kick panel. Restick the edges of carpeting and refasten the splash trim. Make sure there's enough cable in the trunk to reach the antenna, and replace the rear seat.

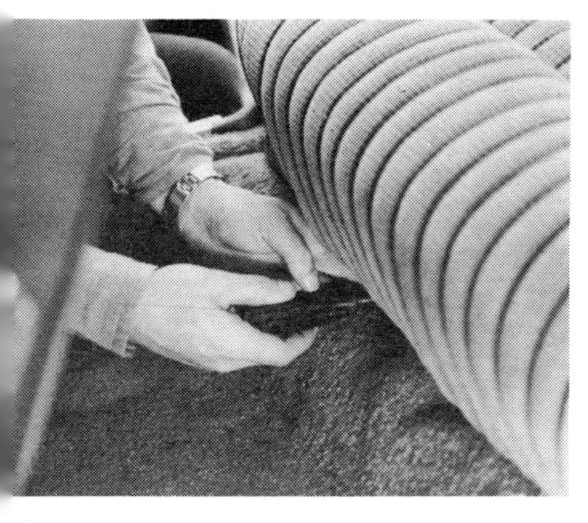

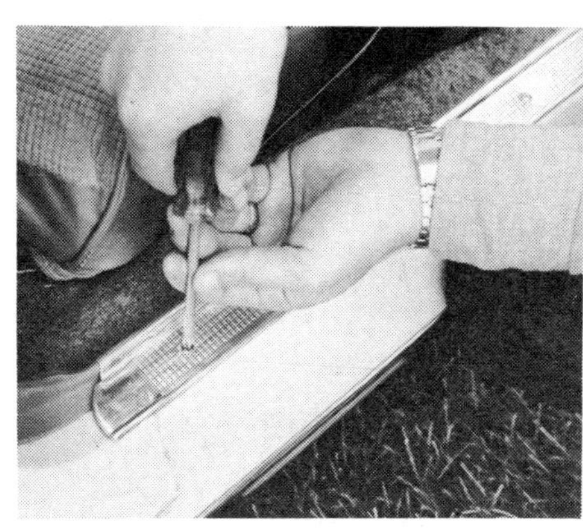

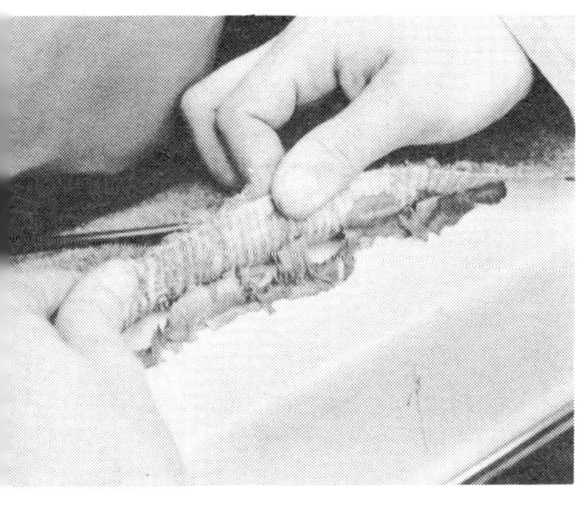

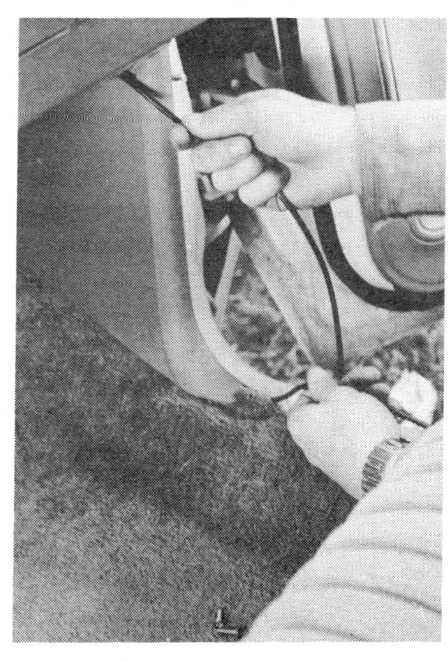

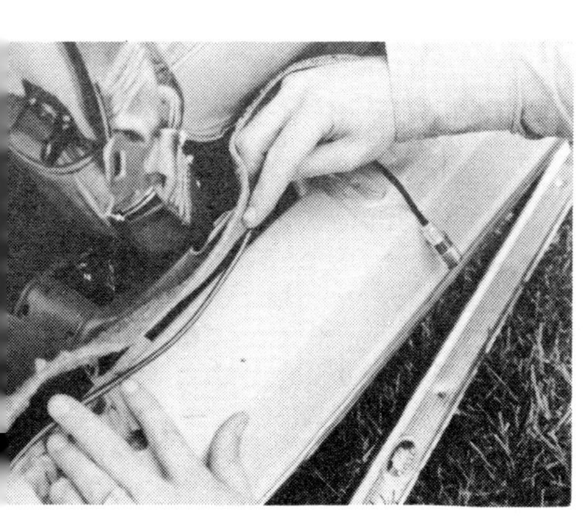

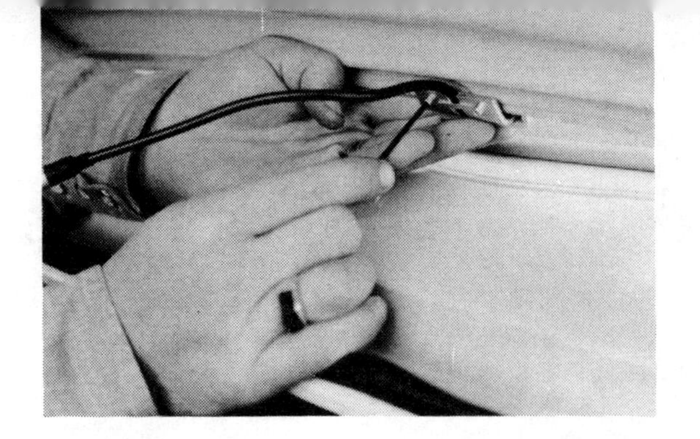

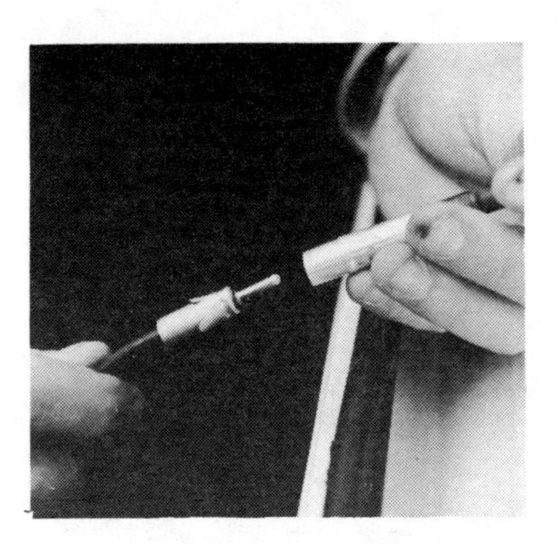

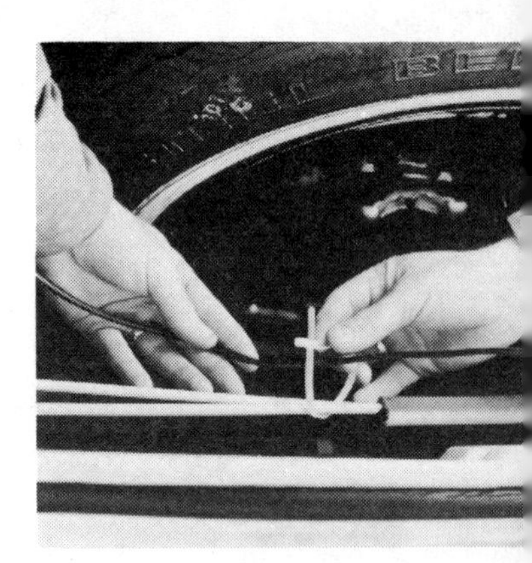

Now, back to the trunk you go. Mount the antenna on one edge of the trunk lid, preferably at the center of the forward edge. Tighten the mounting screws solid, with Allen wrench or screwdriver, whichever the setscrews take. Try grinding them back and forth a few times, too. The object: a thorough electronic ground. Without a firm ground for the antenna mounting base, vswr goes way high. That means you don't get much range with your transceiver.

Push the pin plug into the pin jack, to connect the antenna pigtail to the rest of the lead-in cable. Plug them together firmly, or they won't make connection in the center. That would cause a high reflected reading on the vswr meter, the same as if the antenna were not grounded. Same symptom, different cause.

Finally, install a couple of cable ties to hold the cable up out of the way. You don't want it tangling with whatever you carry in your trunk.

Insert the antenna rod into its mounting spring. Don't tighten the setscrew much yet. You will be sliding the rod up and down a bit to "tune" the antenna.

Connect up the vswr meter. A pigtail or short length of coaxial cable goes between transceiver and meter. The transceiver does not have to be in its mounting bracket, UNLESS it is of the variety that has only one wire to the dc voltage.

Screw the antenna connector onto the ANT end of the meter. Turn on the transceiver. Switch the selector to channel 12. If the channel is clear, key the mike. (If not, ask for a short break, or choose an adjacent channel, or wait a few minutes.) Calibrate the vswr meter on its Forward reading. Then switch the meter to Reflected or Reverse, key again, and note the vswr on the meter scale.

Slip the antenna rod in or out 1/8 inch at a time, keying and reading the meter after each change, to reduce vswr to a minimum. If you can't bring the reading below 1:1.5, you have a serious problem in the antenna installation.

Ideally, if you have trouble, ask a technician to check grounding and continuity with an ohmmeter. If you know how to use an ohmmeter, unscrew the PL-259 connector from the transceiver. You should measure continuity (zero ohms) between the antenna rod and the center pin of the PL-259 rf connector. You should also find continuity between the PL-259 shell and any metal spot on the car. You should find NO continuity (infinite ohms) between antenna rod and car body. If the antenna fails any of these ohmmeter tests, your antenna installation needs a technician.

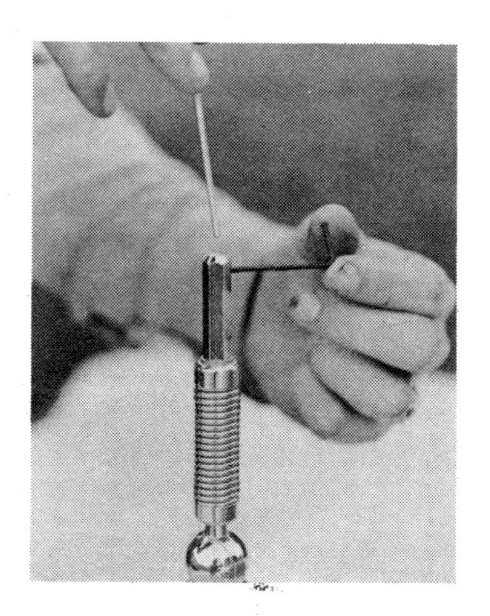

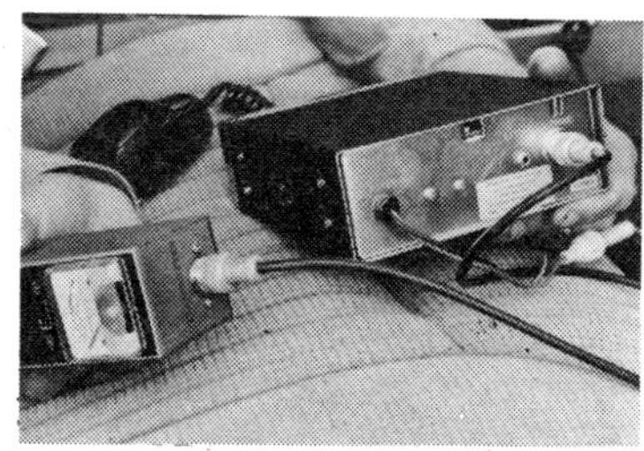

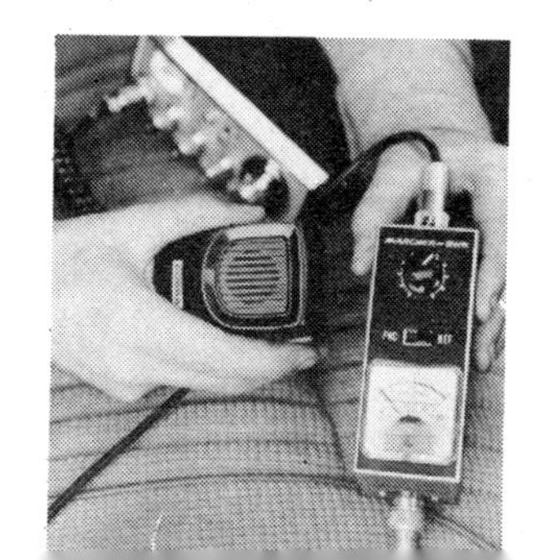

With Squelch open and the car engine running, you'll probably hear a lot of popping noise. In cars with high-energy ignition systems, this interference can really disturb communications.

Noise from spark plugs is hard to combat. Certain other electric devices can be dealt with somewhat, if you want to go to the trouble. Generally speaking, you'd be better off asking an experienced technician for this kind of troubleshooting.

Yet, there are some products on the market that can help in one degree or another. A few are pictured on the page opposite.

At top left is a noise filter for older cars that have dc generators. Not many of those are around any more. But you can use the filter to suppress interference stemming from any dc motor, such as heater, air conditioner, defrost fan, windshield wiper, and so on.

Alternators are more likely to cause noise in CB than dc motors are. Diodes that work in conjunction with alternators can develop a clicking noise in the CB speaker. The clicks are sharper and higher pitched than noise emanating from spark plugs. You connect an alternator filter in the lead that carries alternator current to the voltage regulator and battery.

A noise suppressor kit generally contains a suppressor for the center of the distributor, a suppressor for each spark plug in your engine, and a generator/alternator capacitor (car people call a capacitor a *condenser*). Suppressors affect performance, and may not be advisable for your car. If it's a recent model, ask your dealer if it has "radio" plug wires. They reduce spark-plug interference better than suppressors do, and don't hamper performance so much. You can buy a set, if your car doesn't already have them.

A kit for suppressing stereo *and* CB interference usually deals with noise from the alternator. You hear a variable-pitch whine in the stereo or CB speaker.

Also pictured opposite is a television interference (tvi) filter. Interference filtering has been built into all modern CB designs. But sometimes, usually through tampering, CB transmitters interfere with certain television channels. Rather than spend money on a technician to repair an interfering transmitter, some CBers buy a tvi filter and connect it in the antenna lead-in. That's not curing the defect, but it might keep your neighbors from complaining to the Federal Communications Commission.

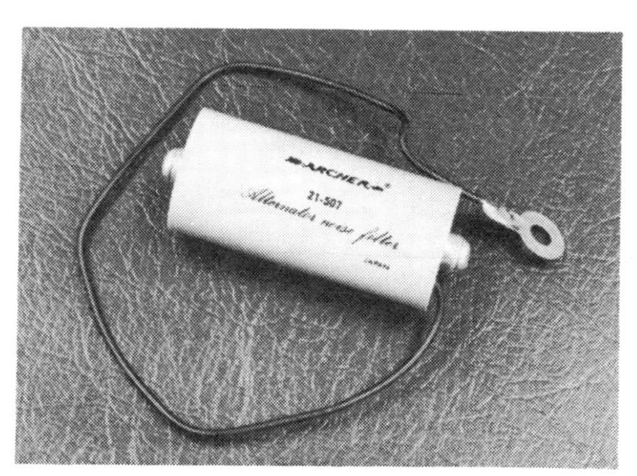
ARCHER
21-507
Alternator noise filter

STOP AUTO RADIO NOISE!!
NOISE SUPPRESSOR KIT
449

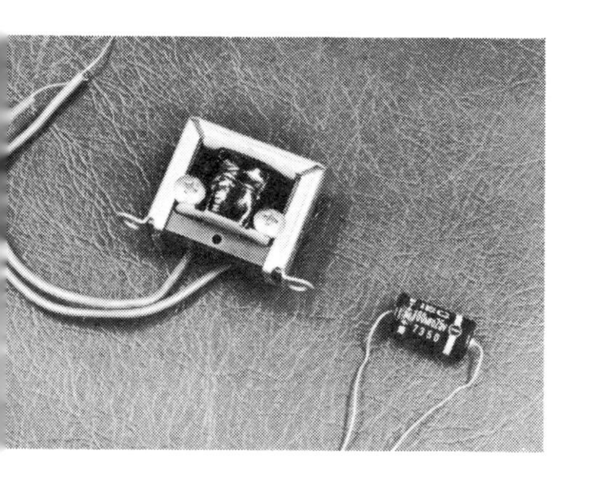

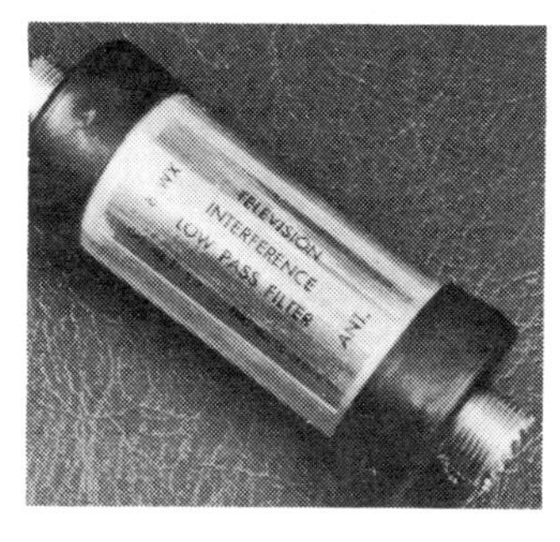
TELEVISION
INTERFERENCE
LOW-PASS FILTER

You have one more thing to do before your mobile installation is truly complete. Tag it. Identify your transmitter with the name and address of the licensee, call sign, and class of station (class D). The licensee (you?) has to sign the ID tag.

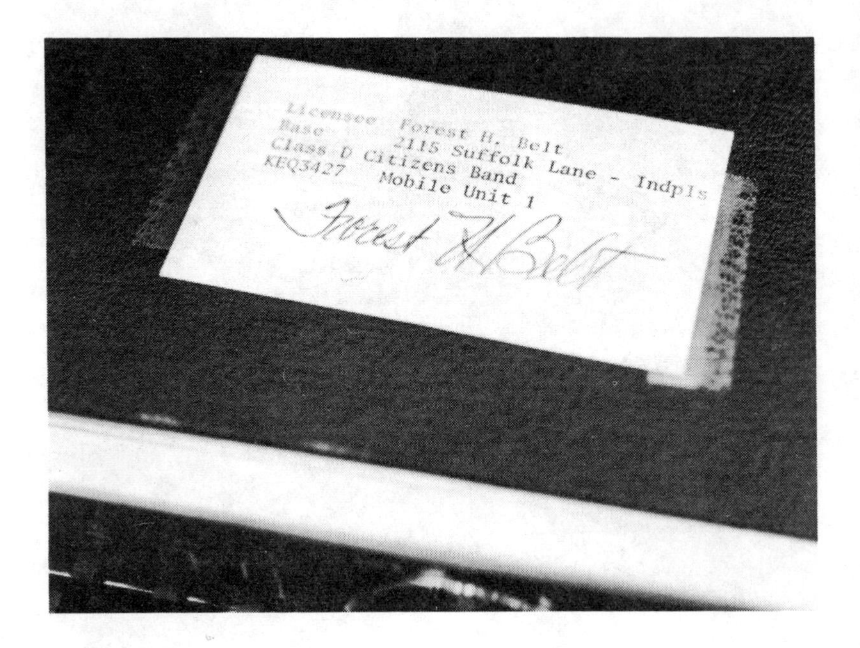

Now you're really ready to CB-it in high style. You can get on those highways with the confidence only awareness can bring. CB keeps you alert, combating any tendency you've had to doze. You won't be surprised, day or night, by traffic jams or accidents.

You'll appreciate this CB radio station in your car more, perhaps, than anywhere else. In a few weeks, you'll wonder how you ever drove without it.

Chapter 7

Recreational CB

Some folks are so addicted to Citizens Band radio that they take their units with them everywhere. Chances to use CB in recreational activities are almost limitless, now that the FCC permits hobby use of the CB channels.

Think of all the ways, and places, you could use CB: on a boat or motorcycle, in a snowmobile, out hunting, fishing, or backpacking. In these last pages of this family CB book, we show you, among other things, how to bring antenna cable into a motorhome neatly. You spend so much for one of those big brutes, you don't want it messed up with a sloppy CB installation. And how you handle the cable often spells the difference.

We even suggest what kinds of transceivers best suit various recreational situations. CB radio today is for fun. Use it that way, to liven up your entire lifestyle.

Maybe you're not a big-time safari hunter, but your campsite could be every bit as elaborate. And that includes two-way radio communications with the outside world. You manage it with the Citizens Band.

Of course you have your mobile transceiver. But operating it much would drain the car battery. Better if you have with you a base station that operates from 117-volt ac power. Or you might bring along a power pack that plugs into 117-volt ac and supplies 12-volt dc: You can take your mobile out and run it from that. Away from power lines, many who camp in motorhomes and trailers carry portable electric generators.

Most versatile of all is the combination base units that can be operated from 117-volt ac *or* from 12-volt dc. If yours can, the owner manual shows how to change it over from one to the other. It either has separate power cords or special terminals on the rear apron. But still, beware of running your vehicle battery down, which could leave you stranded far from home. Operate your base CB only from an auxiliary battery.

CB walkie-talkies lend more freedom to your safari clan. A couple of handheld units, powered from rechargeable NiCad batteries, widen the umbrella of safety that CB affords outdoor people. Anyone who wanders out of earshot should carry a transceiver with one channel active for your "family" channel. Or, if you're camping in a caravan, it should be tuned to the group frequency. You can rotate standing "radio watch" while any campers are away exploring, fishing, or hunting. If everyone goes, carry the most powerful CB walkie-talkie you can afford.

Incidentally, don't count on nonlicensed walkie-talkies. They doom you to communications disappointment. Besides, they're not legal for CB use.

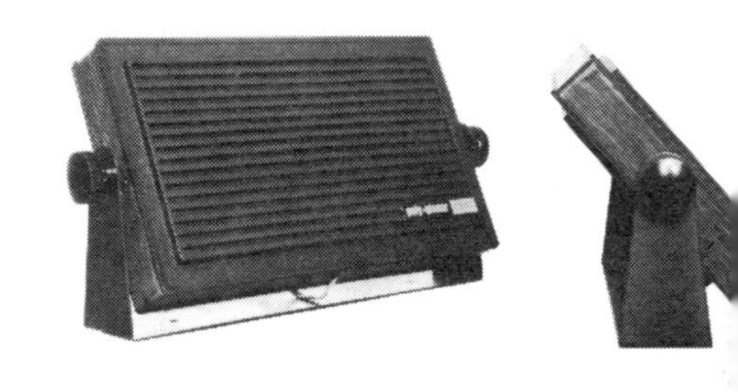

The type of transceiver you own dictates to some extent where it goes in your RV. Or the situation may be the reverse—the amount of space available, or where you want to put it, dictates the kind of equipment you buy. A mobile unit often fits comfortably and handily on or near the driving console. Or, you might hang it overhead, in a pickup camper or minimotorhome. In a trailer or a large RV, you may prefer a base model. Could be that one of those portable units that resemble a tool kit sparks your fancy.

Whatever spot you choose, anchor the transceiver. For example, in some minimotorhomes you find cabinet tops right behind the driver's seat. A base unit left unsecured there during transit could slide off and strike the driver if he had to brake suddenly. So place your unit judiciously and solidly. Make sure no wires or cables dangle to hamper feet or hands. That goes for the mike cord as well.

An auxiliary speaker comes in handy in a motorhome or large travel trailer. One could go at the far end of the interior. You could hook one up outside, for when you're sunning, or relaxing around a campfire. Your local CB shop probably has a variety to choose from. The Polyplanar speaker illustrated here is unique; it's only about an inch thick, and fits conveniently almost anywhere.

Antennas are generally the serious consideration for the CBer who wants to go mobile in a recreational vehicle. Three kinds dominate: rooftop mounts (not too popular), gutter mounted duals (more popular, if the RV has a gutter), and mirror mounted duals (that's the thing). Mirror mounts work almost as well as they do on trucks, since you get that wide spread—up to 102 inches—between the two antennas.

On the road and at distant campsites, range looms important. CB radio does you little good if you can't reach anyone to talk to. So you need distance. And you should keep it "barefoot"—without any linear-amplifier boosting. Distance calls for an efficient antenna system, as high on the vehicle as you can practically mount it.

So we suggest you buy good, strong antennas for your RV, preferably cophased duals. Steer clear of little lightweight models, for they can't give you the reach you want and need. (Some of them don't take the shaking, either.) If you know you can do a right job, install them yourself. But if you're in doubt, hire a good technician. You'll save money in the long run, and you'll have the dependability you count on away from home.

A side-mount like this can be okay. But you've got to be sure you tap into some good solid metal or wood with those mounting bolts or screws. Aluminum skin alone won't do the job. The antenna will be off the first time a limb drags against it. You'll be fortunate if the rip doesn't take some siding with it.

Given solid anchoring, though, an antenna like this works out acceptably. It's high, and height adds miles to your range. If you scrape paint underneath the metal mounting bracket, so there's good metal-to-metal contact, the antenna finds a good ground. As you know by now, that's necessary for efficient radiation and pickup of CB signals. A well mounted antenna, if properly tuned, presents a low standing-wave ratio (vswr). From that standpoint, this antenna operates much the same as the auto antennas described in the preceding chapter.

You might run up against a grounding problem if your RV has fiberglass covering, and several do. Without metal around the base of the antenna, the radiator has nothing to "work against." It's sort of like a reflection in water. Without the metallic "image" of the antenna radiator (the vertical rod), most power sent from the transceiver to the antenna merely gets turned around and reflected back down the cable. That's what creates a high standing-wave ratio. Very little power gets radiated. You can't be heard far. The defect reduces listening range too, but not as drastically.

You can buy what's known as a *grounding kit.* Components in the kit consist of metalic foils that you unroll and attach to the fiberglass skin, but inside of course. Then, in mounting the antenna, you make sure its base contacts the metallic foils. They furnish the "reflection" for the CB antenna. If you were to use dual antennas, one on each side of the vehicle, you'd need two grounding kits. You get even better performance if you manage somehow to run a wire from the grounding foils to the metal chassis or frame of the vehicle. But that's not absolutely necessary.

A couple of manufacturers also make no-ground-plane antennas. These are intended for marine CB installations, because so many boats are wood or fiberglass. NGP antennas provide their own reflections, similar to the way ground-plane base antennas do with their "skirt" of radials. A marine CB antenna has a coaxial skirt, close in around the antenna structure. It occupies far less space than a conventional antenna would. Most important, though, you can find a fairly low vswr despite no metal to "ground to."

How you mount the antenna certainly has a bearing on neatness. But it's really how you handle the lead-in cable that determines whether an installation looks neat or not. It can also determine whether the cable carries water into the vehicle when you travel in rain—or have the outfit scrubbed.

Never, never run the antenna cable in a window. Not only does it look bad, it's hard on the cable to have windows closed on it. Trickles of dampness follow the cable in—and you have enough dampness problems without that. Worse, it leaves the windows open for vandals when you're out of the vehicle for a while. So, when you're tempted to take the easy way, don't.

A properly sealed hole in the side of the vehicle does far better. For one thing you can—and definitely should—put a rubber grommet in whatever holes you drill. You can buy them at electronic stores, or at hardware stores (where they're called *rubber bushings*). You drill a half-inch hole, and put in a grommet, which leaves a quarter-inch hole for the cable to enter by.

A grommet makes the hole easy to seal. You can flow windshield sealer in around the cable. Of course you can do that in a hole without a grommet. But the grommet makes the seal more waterproof and vibration-proof. Besides, a grommet prevents chafing of the cable covering by the sharp metal edge usually left by drilling.

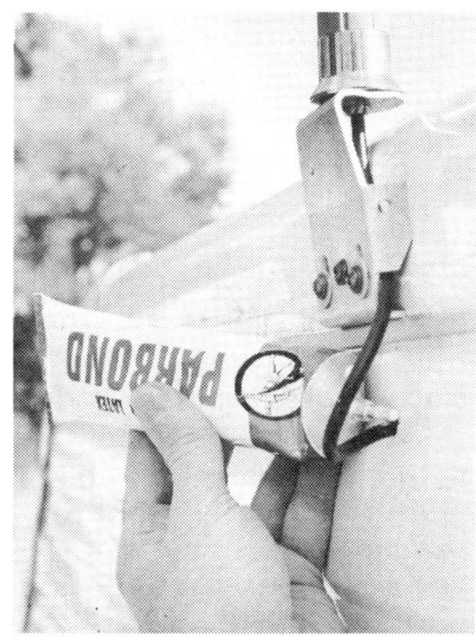

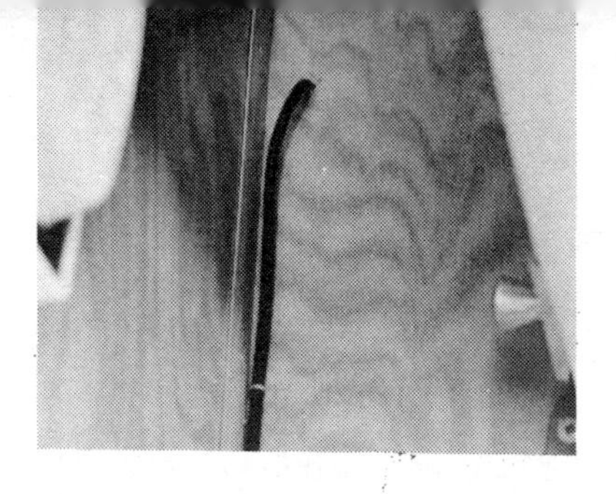

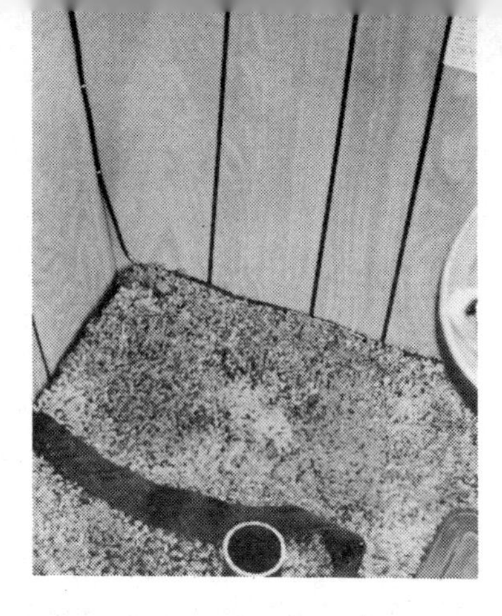

Select carefully where you drill. Always drill from the outside. Drill bits often leave unsightly bulges and dents around the hole, anytime you drill aluminum from the inside out.

Drilling inward, you have to measure carefully to avoid running into partitions. Ideally, you want the drill to come through right beside the bulkhead or partition. Hiding the cable, or at least making it unobtrusive, dresses up any installation. The cable is more easily hidden if it enters the vehicle right beside a bulkhead.

Technicians differ in their methods of handling cable inside an RV. You can't fish it along inside the walls; construction prevents that, once the walls are on (unless you tear paneling off). Typically, you drill your entry hole near the antenna mount. Object: leave as little cable as possible out in the weather. That places the hole high. Inside, you have two choices of direction to run the cable.

In the photo, the technician chose to follow a wall joint downward. A stapling gun, such as phone-company installers use, tacks the cable firmly to the walls. At the floor, the cable then curves (not too sharply, please) and follows the floor/wall juncture forward to the underdash area. The CB transceiver mounts near there.

You can see, however, the technician could just as easily have turned the cable upward and followed the ceiling edge very inconspicuously. Up forward, two or three screws hold in place a cover over the windshield cornerpost. Open that up and you can run the cable down it and behind the instrument cluster to the transceiver. With the overhead bed in storage (running) position, the whole cable run would be virtually invisible. That's the hallmark of a smooth installation.

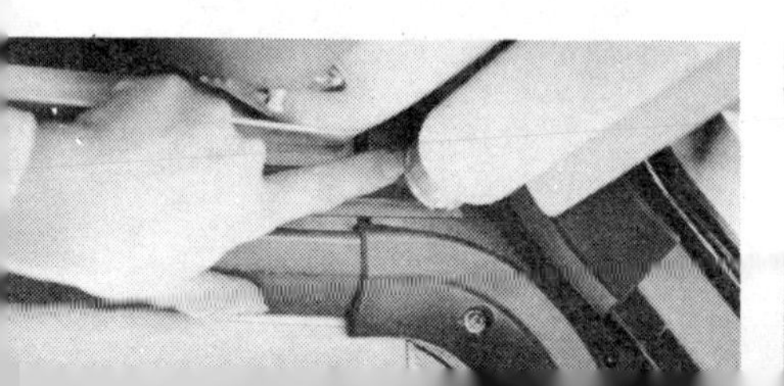

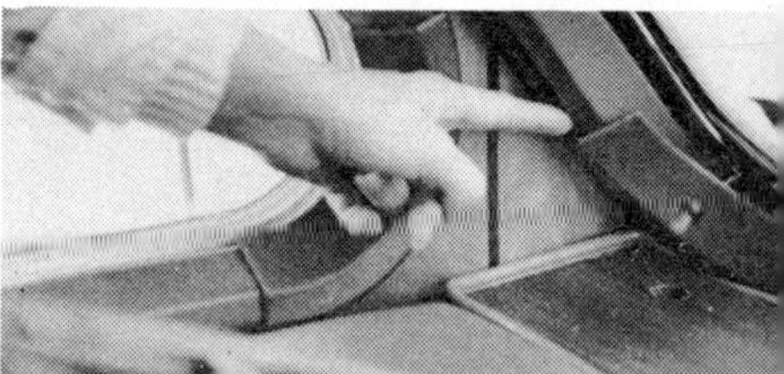

As in automobiles, the 12-volt dc power connection for CB in an RV should be made at the fuse block. Not that the transceiver needs fuse protection; that's already provided (page 98). But the fuse block offers so convenient and accessible a spot to tap into the 12-volt power system. You can use alligator clips if you must, but that's not very permanent. You can find a screw terminal and connect there. By far the best, if you were able to find them, are the quick-disconnect kind of lugs described on page 99.

In most motorhomes, you'll find the power or fuse block just forward of the steering column, and usually a bit to the left. Your RV mechanic can point out the block if it's in some hidden location. He can probably tell you which terminal post is the best to hook onto. He should know which posts are hot with switch on or off, and which ones are hot *only* when you have the ignition switch on. Either is okay for wiring up your CB radio, although you may want to use it many times when the engine is not running. That of course means you must hook to a constantly hot connection point.

You can use the cigarette lighter socket if you wish. That lets you remove the CB unit easily for storage or to take inside when you stay in a motel or park at home.

For many boaters, a CB on board supplements their two-way communication systems. The main radio is a marine or ship-to-shore radio. It's operation is limited to communications with other boats or with marine shore stations. You can only discuss things such as weather reports, right-of-way, emergency calls, and so forth. You need a Marine Radio Service license for the equipment that operates on ship-to-shore frequencies.

Because of Marine Radio restrictions, boaters long ago discovered CB. If you have a very small boat, CB is probably the only two-way radio you bother to have on board. It counts as one of the transmitters authorized by your Citizens Band license.

One primary difference between Marine Radio and Citizens Band radio lies in the frequencies. Marine radio channels sit in a spectrum called *high-band vhf* (very high frequency). CB channels fall at the edge of low-band vhf. Technically, they are hf (high frequency) channels. More important, though, to you at least, is the type of transmissions permitted. On CB, you can discuss the same things with other boaters (if they have CB) that you talk about from home or from your mobile. Moreover, you can call back to your base station or car or camp from the boat. You could not do that easily with Marine radio. Remember, with marine radio you can discuss only boating operations and navigation; you can't tell someone at your base that you'll be docking in about 30 minutes so please toss some fish on the griddle. With marine radio, you can't have a land base (only marine stores and the like can).

Finding a place to put CB radio on a boat isn't always easy or convenient. And it's something you'd better think about before you buy the transceiver.

If you have a canoe or rowboat, you'll want a battery-operated handheld unit. If you have a cabin cruiser, you could set a base unit in the galley, and mount a mobile transceiver near the helm. If you have a small runabout, either inboard or outboard, you may have trouble finding space to hang any unit. Yet, once you comprehend the value of Citizens Band radio, you'll likely find a spot.

Along with your transceiver, the antenna is very important. Most boats, fiberglass and wood ones, lack the metal area to offer the antenna a "ground" to work against. The metal "ground plane" is necessary for effective CB transmission and reception. Install a "no ground plane" antenna. It's what designers call a "coaxial" antenna, and presents a low vswr even without metallic grounding.

Don't forget to seal any holes you make. Windshield sealer prevents leaks and allays water damage to the interior of your boat. The basic installation techniques already shown for the family car or for an RV motorhome guide you sufficiently for installing a CB transceiver on your boat.

One last thing to remember about boat CB. Take your antenna down when you trailer your boat. It only takes a minute or two, and avoids snapping your antenna in two—or ripping it from the boat, with all the damage that entails.

Don't be fooled by walkie-talkies. You'll find in some CB stores a variety of "no license" walkie-talkies. You cannot use them legally to talk to CBers with licensed transceivers. Even though no-license walkie-talkies often hold crystals for several CB channels, they are illegal for communications with anything but other no-license walkie-talkies.

So you ask, "Why are they called CB?" These inexpensive units are very low-powered, putting out no more than 100 milliwatts (thousandths of a watt, abbreviated mW). They are authorized by Part 15 of FCC Rules and Regulations, covering low-power devices that don't need licenses. However, these particular Part 15 transceivers do operate on frequencies within the Citizens Band. Therefore, they can rather loosely be called "CB."

There's no false advertising. You are told that they are low-power no-license transceivers. But the verbalizing still is misleading. Get caught by the FCC using one of these no-license walkie-talkies to talk with a licensed CB station, and you will be fined for operating without a license.

Major differences between no-license 100-mW CB walkie-talkie and class-D handheld transceivers leave the cheapies unsatisfactory anyway, for serious CBing. (1) The 100-mW maximum, a mere 0.1-watt output, has a range of less than a mile, except in rare terrain (from a mountaintop). Licensed sets reach 5, 10, 15, or more miles. (2) Smaller Part 15 walkie-talkie units are really no more than toys. They are sold in pairs as a rule, as they work best in matched sets. Licensable CB outfits are not limited in that manner.

(3) Worst of the traits in no-license sets is their lack of tight frequency control. They "splatter" all around whatever channel they're operating on, thus interfering with more than one CB channel at a time. You'll know what we mean if some neighbor kid gets a pair for Christmas.

Still think you can get by with no-license walkie-talkies? Not if you're a real CBer!

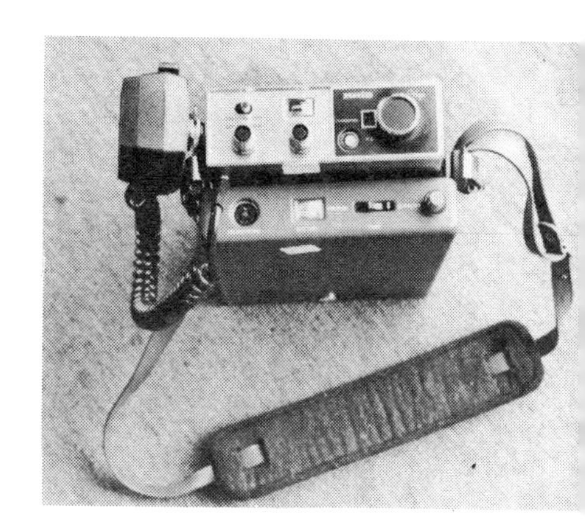

Take your licensed handheld CB along on hunting or backpacking trips. But remember, hilly terrain limits your signal distance, especially when you're in the valleys. Transmitting from a hilltop, of course, your signal goes farther and you receive from greater distances. No hunter or fisherman is fully equipped these days unless he carries a CB walkie-talkie. You can keep in touch with others in your hunting or fishing party, knowing their whereabouts even if you can't see them. We don't need to tell you how handy that transceiver might prove if you get lost.

Backpackers who carry a handheld CB can keep in touch with a base station or camp. They can also alert others to emergencies such as forest or grass fires, landslides, or lost or injured campers. Park rangers often have CB rigs and listen on channel 9.

If you want the versatility of a mobile unit, strap it to a nickel-cadmium battery pack and sling it and a mobile antenna over your shoulder. It's heavy, but you get performance. Some walkie-talkie-type CB transceivers offer the full 4-watts output. Most handheld CB's have less—as little as 1.5 watts. If you're going to be out for an extended period of time, and have a packhorse or a sledge, the power and dependability of a mobile transceiver might prove attractive.

Winter sports can be both fun and hazardous at the same time. Unpredictable weather or a sudden spill in wilderness snow can leave you totally isolated. But if you have CB along, someone may hear your call and send help.

CB in the snow can also be fun. For example, snowmobiling clubs get together sometimes to *foxhunt* with a CB, merely as a weekend diversion.

Foxhunt is a game of reverse hide-n-seek. Hunters use CB to track the Fox. Here's how it works.

You allow your Fox a specific lead time in which to hide. (You may think "fox" tracks easy to follow in the snow. But then you never know how foxy a sly fox can be until you've tried to find him.) Once situated, the Fox must remain at that spot. Then, as he talks, Hunters use the S-meters on their transceivers to fix his location. In other words, you track the Fox by monitoring signal strength. A weakening signal indicates you're moving away from him. Stronger signal, you're getting closer. It's traditional to award a prize to the CBer who pins down the Fox first—perhaps some golden refreshment, a trophy, or an accessory for the CB or snowmobile.

A foxhunt can be more than just fun. It makes a good CB-rescue training exercise. Hunters have to move the "injured" Fox back to base camp. You can use these signal-finding techniques to locate really lost snowmobilers, skiers, backpackers, and so on.

CB radios on motorcycles are not uncommon. In fact, some CB shops carry kits specifically for motorcycle installation.

One critical point to remember with motorcycle CB: Dangling cables anywhere spell BIG trouble. The only right way to run coax from antenna to transceiver is under the seat and across the gas tank, and up to the instrument cluster. Don't let anything crimp or pinch the cable, anywhere. Tape or cable-tie the coax along its path from antenna to transceiver so it cannot slip out of place and become dangerous. Loose wires also wear faster and will cause communication problems.

You might be surprised to learn that you can hear a normal CB transceiver at speeds up to 60 mph, without an earpiece or any extra speaker. A windshield helps. However, if you prefer, you can listen through an earphone. Just plug a taperecorder-type earphone into the auxiliary speaker jack. Get one that pushes into your ear but does not clip on your ear or head; the latter could break your neck in a spill.

As for getting your transceiver stolen, small sets like this Cobra 19 fit neatly into one of your lock compartments. Out of sight, out of (some thief's) mind. These compartments are relatively sturdy and not easily broken into. Most thieves find it safer to rip off an auto installation; a cyclist might carry an extra chain for beating CB thieves and the like.

When you're having fun, time passes fast. Your license is good for five years. But you may be astounded at how quickly that much time slips away.

Sometime before your CB license runs out, say three or four months, obtain a new Form 505. Fill it in, mark the Renewal box, and ship the application off to the Gettysburg address (page 47). Renewal costs another $4. Renewal may take that whole three months, depending on how backlogged the Commission office in Gettysburg is. Actually, you can apply for a renewal anytime during the final year of your old license.

Just don't forget. You'll be operating illegally if you don't have the new license back before the old one expires. Get the job done on time, and you'll be in business for another five years.

On the lines below, record the model and serial numbers of all your Citizens Band radio equipment. If any of your units, such as mikes, have no serial numbers, scratch your Social Security number on it with a sharp knife or an engraving tool. Record that. Then, if your equipment gets stolen, you have this record for insurance purposes and for the police.

Take a photograph of your equipment, with Instamatic or Polaroid if that's what you have. Tape that photograph to this page; it and the serial numbers can help you prove your insurance claim, if your equipment ever does get stolen.

ANTICRIME REGISTRATION FOR CB RADIOS

Base Station	Brand	Model No.	Serial No.
Base transceiver			
Standby transceiver			
Extra transceiver			
Microphone			
Beam antenna			
Antenna rotator			
Monitor antenna			
SWR Meter			
Modulation monitor			
Other (describe)			

Mobile	Car Make	Brand	Model No.	Serial No.
Mobile 1				
Mobile 2				
Mobile 3				
Mobile 4				
Mobile 5				
Other (describe)				

Handheld	Brand	Model No.	Serial No.
Unit 1			
Unit 2			
Unit 3			
Other equipment (describe)			

Date____________________ Signed____

You have in this book all the help you need to become an active and legitimate CBer. And once you get started, don't blame us if you become addicted. We warn you: CB is contagious. You won't get as many books read as you used to, nor see as many television shows. You may even forego the Saturday night bridge or poker game, at least some weekends.

More than anything else, though, even if you don't become a CB nut, you'll find Citizens Band radio turning more and more into an indispensable tool in your everyday moving around (or even in your daily staying-at-home). CB radio will save you time, it may save you money, and it could upon occasion save your life or that of someone you love.

If that sounds as if we're promoting CB, it's because we're sold on its usefulness ourselves. Try it. You soon will see what we mean. Enjoy CB. That's what it's for.

Seventy-threes to you, and all those good numbers. We'll be listening for you when we travel through your area, or when you travel in ours. This is KEQ3427 on the side. We're down.